Carl Friedlaender, Stephen Yates Howell

A manual of microscopical technology

For use in the investigations of medicine and pathological anatomy

Carl Friedlaender, Stephen Yates Howell

A manual of microscopical technology
For use in the investigations of medicine and pathological anatomy

ISBN/EAN: 9783743357457

Hergestellt in Europa, USA, Kanada, Australien, Japan

Cover: Foto ©berggeist007 / pixelio.de

Manufactured and distributed by brebook publishing software (www.brebook.com)

Carl Friedlaender, Stephen Yates Howell

A manual of microscopical technology

A MANUAL

OF

MICROSCOPICAL TECHNOLOGY

FOR USE IN

THE INVESTIGATIONS OF MEDICINE AND PATHOLOGICAL ANATOMY

By Dr. CARL FRIEDLÆNDER

LECTURER ON PATHOLOGICAL ANATOMY IN THE UNIVERSITY OF BERLIN

TRANSLATED, WITH THE EXPRESS PERMISSION OF THE AUTHOR, FROM THE SECOND
ENLARGED AND CORRECTED EDITION

By STEPHEN YATES HOWELL, M.A., M.D.

BUFFALO, N. Y.

NEW YORK & LONDON

G. P. PUTNAM'S SONS

The Knickerbocker Press

1885

Press of
G. P. PUTNAM'S SONS
New York

IN presenting this version of Dr. Friedlaender's valuable treatise to the American student and practitioner, I deem it advisable to add a word in explanation of what I have endeavored to accomplish.

The translation was entered upon in 1884, but from various causes its completion has been delayed until the present time. However, by thus devoting to the work a much greater amount of time than would have been required for a merely literal rendering of the original, I have been enabled to carefully elaborate portions of certain chapters, where such elaboration seemed likely to enhance the practical value of the manual in its new field.

Chapter I., treating of the stand and its various accessories, has been largely rewritten, and has been completed by the addition of a list of American and foreign manufacturers, which includes the descriptions and prices of their latest complete student microscopes. An illustrated article on the comma-bacillus of Asiatic cholera, treating of the morphology, cultivation, staining, and examination of this specific microbe, has also been inserted, and will be found to embrace the results of the most recent investigations of this all-important subject. And, finally, numerous foot-notes have been added throughout the work, which were planned to render as clear and comprehensive as possible, the various views and suggestions of the author.

I have thus laid stress upon some of the distinctive features of my work ; since, while it has been passing through the press, my attention has been called to another translation of Dr. Friedlaender's volume, which has just appeared, and which, though begun at a later date, was apparently prepared in ignorance of the fact that I had a translation under way. Without attempting any criticism of this competing edition, it is not out of order, I think, to point out the fact that it represents only a literal version ; the translator not having thought it necessary, as I have done, to add any material or notes to the original work, either for the purpose of considering one of the most absorbing topics of the present day, or of adapting it to the special requirements of the American student.

S. Y. H.

164 Franklin Street, BUFFALO, N. Y.,
October, 1885.

AUTHOR'S PREFACE.

THE author has frequently been requested to write a short description of the methods employed in microscopical investigations of a diagnostic or pathological nature ; and this little work is the outcome of a determination to accede to the wishes of his friends. The subject, which was an extremely simple one as late as the seventh decade of the present century, has gradually grown more complicated, and, in many respects, has undergone very marked improvement and refinement.

A large share of the surprising progress made during the last few years, particularly in the field of vegetable parasites, results directly from these improvements in technology.

If, then, every one, who wishes simply to comprehend the advance of pathology, must learn the later methods, this is especially true of one who desires, personally, to make microscopical examinations for professional purposes.

Thus far a comprehensive exposition of the recent improvements in the field of pathological investigation has been wanting, though painfully missed ; and this little volume tends to fill up the gap.

The relatively greater attention given to the articles on schizomycetes (bacteria) will surely meet with general approval.

In several instances the diagnostic and prognostic significance of phenomena might have been further elaborated ; as has been done in the consideration of the bacilli found in the sputa of tubercular patients, and in differentiating between erosion and carcinoma of the cervix uteri.

May the little book fulfil its mission, and prove a trusty guide to the student, who is beginning his studies in a field as enticing as it is difficult ; and, it may be, the more experienced investigator will also find a useful hint, here and there.

CARL FRIEDLAENDER.

City Hospital, Berlin,
August, 1882.

PREFACE TO SECOND EDITION.

AFTER the lapse of a comparatively short time, a second edition of this little work has become necessary ; and I have assumed the labor of searching out and adding to the text such new matter, as would really tend to enhance its utility.

Besides this, and in response to the expressed wishes of many, I have added a colored plate, illustrating the most important and characteristic pathogenic schizomycetes. For the drawings, I am indebted to my esteemed friend and co-worker, Dr. Gram, of Copenhagen. They all represent an enlargement of 1,000 diameters.

THE AUTHOR.

BERLIN, *April*, 1884.

CONTENTS.

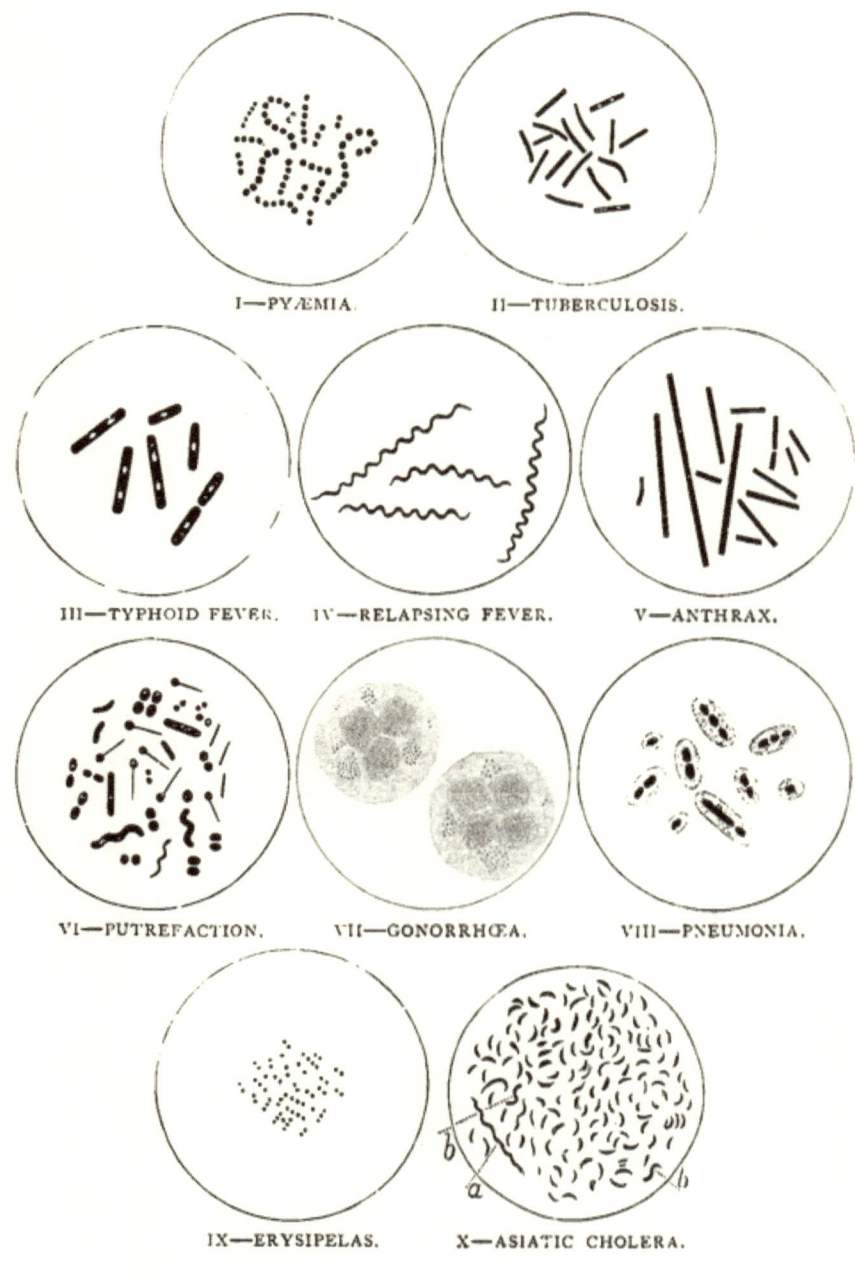

I—PYÆMIA. II—TUBERCULOSIS.

III—TYPHOID FEVER. IV—RELAPSING FEVER. V—ANTHRAX.

VI—PUTREFACTION. VII—GONORRHŒA. VIII—PNEUMONIA.

IX—ERYSIPELAS. X—ASIATIC CHOLERA.

EXPLANATION OF PLATE.

Fig. X—Represents an enlargement of 600, the others of 1,000, diameters.

Fig. I—Rows or chains of micrococci, from pus.

Fig. II—Tubercular bacilli, from the juice of a miliary tubercle.

Fig. III—Bacilli of typhoid fever, from a Peyer's patch.

Fig. IV—Spirilla, from the blood of relapsing fever.

Fig. V—Small rods and filaments, from the blood of anthrax.

Fig. VI—Various forms of bacteria found in the saliva.

Fig. VII—Pus-corpuscles, with aggregations of micrococci, from the secretion of gonorrhœa.

Fig. VIII—Capsulated micrococci of pneumonia.

Fig. IX—Micrococci of erysipelas, from a section of the cutis.

Fig. X—Comma-bacilli of Asiatic cholera. Object-glass preparation, from bouillon cultivation. *a*, Long, screw-shaped threads ; *b*, the *S*-shaped forms.

MICROSCOPICAL TECHNOLOGY.

I. THE MICROSCOPE.

THE principal rule to be observed in the selection of a microscope, is, that the lenses and stand should be fault-less. One should not permit a somewhat lower price to betray him into the purchase of an instrument of inferior quality; for, the majority of the objects which we are called upon to examine are so difficult, from their deli-cacy of contour, etc., that only under the most favorable conditions do we succeed in bringing them into view in a satisfactory way, and without great loss of time.

Accordingly, one should, at the start, select an instru-ment from a well-known, reliable house. In case the lenses, or other portions of the apparatus, do not satisfy *all* reasonable demands, they should be immediately returned; as nothing is more unpleasant than a lasting struggle with an imperfect microscope. But, on the other hand, it is by no means necessary to purchase, at the outset, the most powerful objectives, as these, in particular, enhance the price of the instrument ; but the beginner should rather employ lenses of low and medium power ; *i. e.* with an amplifying capacity not to exceed 300 diameters. The manipulation of stronger objectives is attended with so many difficulties, and demands such a painful degree of accuracy, that a preliminary training in the practical use of moderate powers, for a considerable period, is decidedly advisable.

I.—The Stand.   Abbé's Apparatus.

The stand must be so arranged, that objectives of the highest power may be employed ; the stage should be large, solid, stiff, and provided with a roomy aperture, through which, the diaphragm having been removed, a transverse section of the spinal cord, for example, could be examined *in toto* under a low power ; and, in particular, the movement of the fine-adjustment screw should be sufficiently delicate.   As a rule, provision for the inclination of the body is superfluous.

The cylindrical diaphragms—the revolving variety is less perfect—should, as a matter of course, be accurately centered and easily changed.   When inspecting unstained objects with lenses of considerable magnifying power, diaphragms with small openings are employed, in order to secure sharp definition; while, for low powers, a wide-apertured diaphragm is usually necessary for the complete inspection of the field. In all cases, a condenser, or the Abbé apparatus, is desirable ; while, indeed, for the study of bacteria,* it becomes a necessity.

The rays of light, which are reflected by the mirror upon the lens of the condenser, are so refracted by it, that they all come together at one point, the focus, which exactly coincides with the position of the object. In this way the object receives from below an enormous amount of light,—not, indeed, a bundle of approximately parallel rays merely, as in the usual examination with a diaphragm of small aperture, but a complete cone of the greatest possible angle of aperture, at the apex of which the object is placed. Thereby the fine contours of a transparent object, as far as they depend upon differences in refractive power, are wellnigh completely lost,— wiped out, as Koch expresses it.

The colored portions of the object, however, which were before partially or wholly masked by the contours of the colorless parts, stand out all the sharper (Isolirung

---

* Prof. Nägeli, of Munich, in his work upon " The Inferior Fungi, and their Rôle in Infectious Maladies," separates those varieties which produce decompositions into three groups : first, the Mucorini, or mould fungi ; second, the Saccharomycetes, or budding fungi, which produce the fermentation of wine, beer, etc.; and, third, the Schizomycetes, or fission fungi, which produce putrefactive processes. As the last correspond to our *Bacteria*, I shall hereafter substitute the latter popular, and more familiar, term for the *Schizomycetes* of the original text.—S. Y. H.

des Farbenbildes—Koch) ; and, hence, by examining with the open condenser, it is often possible to recognize deeply stained micro-organisms or other minute colored bodies, which, with the ordinary illumination, are obscured by the structural details of the object, and appear indistinct, or even irrecognizable.

The Abbé condenser has an angular aperture of 120° ; while the older forms have usually a much smaller angle, and, hence, prove unsatisfactory. Under the lens of the condenser is a carrier, into which may be inserted diaphragms, having apertures of various sizes and shapes. When a diaphragm of small aperture is employed, the illumination is, naturally, quite similar to that obtained through a cylindrical diaphragm having a small opening. When no diaphragm is used, the greatest volume of light is obtained, and the condenser is said to be *open.* The remaining complicated appointments of the Abbé apparatus are, thus far, of little value for our purposes. The use, then, of the illumination afforded by the open condenser in the examination of all stained preparations, is attended with great profit ; indeed, it becomes indispensable in many difficult investigations ; and we owe much to Koch for its introduction.*

In selecting a stand, therefore, care should be taken to

* R. Koch : " Untersuchungen für Aetiologie der Wundinfectionskrankheiten. Leipzig, 1878.

see that a well-made condenser of large angular aperture, or an Abbé apparatus, is, or at least can be, adjusted to it. The beginner, however, can do perfectly well without this accessory, as, practically, it is only employed for the finer investigations with high powers.*

2.—OBJECTIVES.

In selecting these most important adjuncts of the microscope, one should obtain :

1. An objective of low power, having an equivalent focus of one inch, and affording an enlargement of about 30 diameters with a medium eye-piece,—to be used for the inspection of large sections from the brain, cord, liver, and kidneys ; in searching for trichinæ, etc.

2. One of medium low power, or $\frac{1}{2}$ inch equivalent focus, and magnifying from 60 to 80 diameters ;

3. A moderately high power, having an equivalent focus of $\frac{1}{5}$ inch, with which an enlargement of 400 diameters may be obtained ; and,

---

* Among the many and excellent sub-stage condensers, manufactured in this country, none are more worthy of special mention, both for general efficiency and moderate price, than Prof. E. Abbé's New Illuminating Apparatus, as constructed by J. Grunow, of New York City, and that supplied by the Bausch and Lomb Optical Co., of Rochester, N. Y. Both may be employed either dry or with water-immersion, and are furnished in similar mountings for $22.00. By employing an *adapter*, either with or without a centering adjustment, objectives may be made to serve as condensers with good effect ; though, as a matter of course, their range is comparatively limited.— S. Y. H.

4. A quite powerful immersion lens, equivalent focus $\frac{1}{12}-\frac{1}{16}$ inch,—to be employed in investigations of a finer sort.*

Of these, Nos. 2 and 3 are most used in our work.

Immersion objectives are employed when strong magnifying power is a desideratum, and their proper manipulation requires a certain amount of care and practice ; as we said before, the beginner had better do without them for a time.

A complete elucidation of the advantages obtained by immersion, would call for a more or less elaborate discussion of light, and the laws which govern it ; but the following will, probably, suffice for our purpose :—

In passing through a cover-glass, the perpendicular, or axial, ray of the pencil, which diverges from any point of an object beneath, is not deflected from its course ; while oblique rays are refracted to a degree proportional to their primary deviation from the axial ray, and, when traced backwards, seem to emanate from points which successively approach the under surface of the cover.

---

*The permanent objective-systems are variously designated. On the Continent, the opticians usually employ arbitrary marks,—as numerals, which increase with the strength of the combination, or letters, variously arranged ; while in England and America, the objectives, both dry and immersion, are rated according to the focal distances of simple lenses of like strength. Thus a ¼-inch combination of lenses, or objective, corresponds in strength with a simple lens having a like focal distance.—*S. Y. H.*

This apparent upward displacement of the object, which is called *negative aberration,* thus increases with the obliquity of the rays, and, as will readily be appreciated, affects the precision of the image formed by an objective more and more, as its capacity for transmitting oblique rays is increased.

If, now, we call the angle, formed by the most divergent rays which can pass from any point of an object through the front lens of an objective and take part in the formation of an image, the *angle of aperture* of that objective, it is clear that, in order to retain the distinctness of the image, this angle must either be very small, or the "negative aberration" must be corrected. Low and medium powers, therefore, are perfectly corrected, at the outset, for covers of average thickness, and even objectives of high power and large angular aperture (focus $\frac{1}{12}$ inch, ang. aper. 130°) are made to perform excellently, by correcting them for cover-glasses of a standard thickness, supplied by their makers; though, for objectives of less than $\frac{1}{8}$-inch focus, the immersion system is preferable. In objectives of the best quality, however, the component lens-systems are so mounted, that, by turning a graduated screw-collar, a rectilinear movement is imparted to the posterior systems, whereby their distance from the end, or front, lens is varied ; and, as opposite aberrations have been given to the former and latter systems, by over- and under-correcting them, such an excess of positive aberration may easily be given to the whole combination by approximation, as shall neutralize the negative aberration occasioned by the cover-glass.

However, the necessity for this accurate cover-correction is greatly lessened by the introduction of a layer of *water* between the cover-glass and the front lens of the objective (Water-immersion) ; as the rays are refracted less in passing from the cover-glass into that fluid than into air, and the negative aberration is correspondingly diminished.

Then, too, rays of the same obliquity suffer much less loss by reflection, in passing into the marginal portion of the lens from water.   Hence, the advantages in favor of water immersion—the angular aperture being the same—are, lessened negative aberration, increased light, and greater working distance between the objective and the object.

But these advantages are all increased, if we substitute, for the water, a fluid having the same refractive and dispersive power as crown-glass ; in which case, even the most oblique rays pass from the cover to the convex back-surface of the front lens of the objective without refraction, dispersion, or any material reflection.   Such an arrangement is called *homogeneous immersion*, or, as oil is the medium generally employed, *oil immersion.* The fluids used are oil of cedar-wood, or a mixture of oil of fennel and castor oil ; while, lately, a solution of chloral hydrate in glycerine has been recommended. This system of "homogeneous immersion" enables us to realize all the advantages due to large angular aperture, such as increased brilliancy of the image and greater resolving power ; while the objectives themselves are the master-pieces of our opticians, and shed lustre upon the names of Abbé, Zeiss, and Stephenson, to whose labors they chiefly owe their introduction. Though usually deemed superfluous, adjustable mountings should be employed for these objectives, when exactitude is required ; inasmuch as variations in the refraction of cover-glasses, in the immersion fluid, in the length of tube, as well as change of eye-pieces, etc., seriously affect their usefulness. The use of an oil-immersion objective is attended with difficulties, even in the hands of an expert.

When employing a water-immersion lens, one soon learns the proper size of the drop of distilled water, which is to be placed upon the cover-glass, or front lens of the objective, with a glass rod or camel's-hair brush ; and he should take care that it does not flow over the edge of the cover, and mingle with the fluid underneath —an accident which cannot, of course, occur in the case of preparations already mounted for preservation. A fine capillary tube will remove the water-drop from the cover ; while a vigorous wiping is required to free its surface from oil.   For the most part, however, it suffices to remove only the superfluous oil ; and this may be accomplished by passing fine blotting-paper lightly over the surface of the cover-glass.

3.—Eye-pieces and other Accessories.

Two Eye-pieces are generally employed : a *shallow*, or weak, one, for ordinary use ; and a *deep*, or more powerful, one, for special cases.   An eye-piece micrometer should be fitted for use with either eye-piece.

Among the accessories, supplied by dealers in microscopical apparatus, are :—

1. The Nose-piece, or Revolver, for the rapid change of objectives.*

2. A Drawing Apparatus, or Camera Lucida.   That known as Oberhäuser's is one of the best.†

---

* M. Nachet's *porte-objectif*, or objective-carrier, promises well as a substitute for the nose-piece, it being simple, effective, and of moderate cost.—S. Y. H.

† Excellent prism Camera Lucidas are supplied by all leading opticians, and Grunow's new model may be mentioned as an admirable type of this class. However, where the price of such instruments cannot be afforded, Beale's *Neutral Tint Reflector*, which is both efficient and cheap, may be commended ; or the microscopist may easily make for himself a useful modification of Beale's reflector, by taking a flat cork about $1\frac{1}{4}$ inches in diameter, cutting a hole through it large

3. A Polarizing Apparatus.

4. A Micro-spectroscope.

5. A Warming-stage, either Schultze's or Stricker's.

6. A Stage Micrometer.

Complete Student Microscopes.

For the convenience of the reader we shall now mention a few approved combinations, of moderate price, which, as they suffice for general histological work, for the examination of the urine, etc., will be most apt to meet with the requirements of both students and practitioners of limited means. The method by which coarse adjustment is effected,—whether it be by sliding the main tube through a sleeve, by imparting a spiral motion to it, or by means of the rack and pinion,—is given in each instance.

The Bausch and Lomb Optical Co., Rochester, N. Y. (1885) :

Harvard Microscope (sliding tube), with two eye-pieces (A and C), and two objectives (1 in. and ¼ in.), in an upright, or horizontal, polished case, $45.00.

Model Microscope (rack and pinion), with one eye-piece, two objectives (1 in. and ¼ in.), camera lucida,

---

enough to enable it to replace the cap of the eye-piece, and making a transverse slit beneath the aperture, which shall receive a thin cover-glass at an angle of 45° with the axis of the cork.—S. Y. H.

pliers, slides, and covers, in a completely furnished, upright case, $45.00.

Joseph Zentmayer, 147 South Fourth Street, Philadelphia (1884) :

American Student Stand (sliding tube), with one eye-piece (A or B), two objectives ($\frac{8}{10}$ in., ang. aper. 24°, and $\frac{1}{8}$ in., ang. aper. 75°), in a walnut case with lock and handle, $38.00.

W. H. Walmsley & Co., 1016 Chestnut Street, Philadelphia (1885), agents for R. & J. Beck, of London :

The *Monocular Economic* Microscope (sliding tube), with one eye-piece, two objectives (1 in. and $\frac{1}{4}$ in.), side condensing-lens, glass plate with ledge, pliers, etc., in a polished case, $45.00.

Benj. Pike's Son & Co., 930 Broadway, N. Y. (1884) :

The *Professional* Microscope (rack and pinion), with two eye-pieces, two objectives ($\frac{3}{4}$ in. and $\frac{1}{8}$ in.), bull's-eye condenser on a separate stand, forceps, etc., in a mahogany case, $50.00.

T. H. McAllister, 49 Nassau Street, N. Y. (1884) :

Student's Microscope (rack and pinion) with one eye-piece, two objectives (1 in. and $\frac{1}{8}$ in.), walnut case, $42.00.

James W. Queen & Co., 924 Chestnut Street, Philadelphia (1884), agents for Henry Crouch, of London, Eng., and John W. Sidle & Co., of Lancaster, Pa. :

Crouch's New Histological Microscope (rack and pinion), with two eye-pieces, two objectives (1 in. and $\frac{1}{4}$ in.), glass slip with ledge, in a mahogany case, $55.00.

John W. Sidle & Co.'s *Acme* Microscope, No. 4 (rack and pinion), with two eye-pieces, two objectives ($\frac{3}{4}$ in. and $\frac{1}{4}$ in.), in walnut case, $55.00.

J. Grunow, 70 West 39th Street, N. Y. (1884) :

Student's Microscope, No. 1 (sliding tube), with one eye-piece, one objective ($\frac{1}{4}$ in., 100° ang. aper.), in case, $45.00.

Herbert R. Spencer & Co., Geneva, N. Y. (1885) :

The *Nonpareil*, No. 4 (rack and pinion), with one eye-piece, two objectives (1 in. and $\frac{1}{4}$ in.), in a fine walnut box, $49.00.

Charles X. Dalton, 40 Hanover Street (room 31), Boston, Mass. (1884), successor to R. B. Tolles :

Tolles' Student's Microscope (sliding tube), with one eye-piece, two objectives (1 in. and $\frac{1}{4}$ in.), in a black-walnut case, $50.00.

Miller Brothers, 69 Nassau Street, and 1213 Broadway, N. Y. (1884) :

Student's Microscope (spiral motion), with one eye-piece, two objectives ($\frac{2}{3}$ in. and $\frac{1}{4}$ in.), glass stage, in a black-walnut case, $50.00.

Walter H. Bulloch, 99 and 101 West Monroe Street, Chicago, Ill. (1884) :

New Student Stand (rack and pinion), with one eye-piece, two objectives ($\frac{3}{4}$ in. and $\frac{1}{4}$ in.), in a case, $50.00.

When desired, a sub-stage condenser can be fitted to any of the above stands. The illustrated catalogues of the various manufacturers of optical instruments will prove of great assistance in the selection of a microscope adapted to one's individual requirements and purse ; though the inexperienced are earnestly advised to avail themselves of the opinions of competent judges before purchasing.

Of the foreign manufacturers of microscopes and microscopical accessories, we may mention Powell & Lealand, Ross & Co., C. Baker, and Charles Collins, of London ; Nachet, Verick, Chevalier, and Prazmowsky, of Paris ; L. Bénèche, and Schmidt & Haensch, of Berlin ; Hartnack, of Potsdam ; Zeiss, of Jena ; Winkel, of Göttingen ; Leitz, and Seibert & Kraft, of Wetzlar ; and Merz, of Munich.

II.—OTHER UTENSILS.

1.—LAMPS.

DAYLIGHT affords the best illumination, especially that reflected from a white cloud in the neighborhood of the sun. Direct sunlight cannot be employed ; yet it is always most advantageous to place the work-table before a window which has a southern outlook. When the sun shines, part of the window may be screened with a white shade and the work continued,—the light being taken from the shade, or from the sky through the unscreened portion of the window.

On foggy and dark days, however, a lamp becomes necessary, especially when high-power objectives are used ; while, for evening work, it is indispensable. For this purpose we make use of an argand gas-burner, to which a chimney of isinglass is fitted, together with a shade made of stiff paper.

The direct light of the flame is employed, its yellow color being corrected by a disk of blue glass, which fits into a ring above the eye-piece. By having several such disks of various shades of blue, proper corrections can be made for the various positions in which the lamp may be placed. The burner is attached to a stand by a movable arm, and is fixed only 20–30 cm. above the level of the table. In case the screen is properly attached to the burner, one need scarcely notice the heat of the flame ; and, thus, the work is rendered hardly more trying to the eyes, than when daylight is used.

A kerosene lamp, fitted with a good-sized round burner, or one of the duplex variety, also affords an excellent light.*

---

* A small, flat-wick lamp answers very well, particularly when a bull's-eye condenser is available ; though the stronger light afforded by the German Student lamp, Hitchcock's mechanical lamp—which

To be recommended, is the employment of a glass globe some six inches in diameter (Schusterkugel), which is filled with a solution of the ammonio-sulphate of copper, and placed between the lamp and mirror of the microscope. A few drops of ammonia are added to a solution of copper sulphate, when a fine blue color appears. If more water be afterwards added, a cloudiness usually ensues, which is dissipated by the further addition of ammonia. The proper intensity of color can readily be ascertained by testing ; when a beautiful white light is obtained, which falls in parallel rays upon the surface of the mirror. The globe is to stand low ; and the light should be employed for the microscope only, not for the preparation of specimens, etc.

2.—Glass Apparatus.

The slides, or object-glasses, should be made of the best white glass, of uniform thickness and shape, the edges being ground smooth.

For general use, a slip of glass three inches long and one inch wide is employed.

Cover-glasses, likewise, are of a definite, medium

---

dispenses with a chimney,—and Fiddian's portable illuminator, offers some advantages. In any case, however, great care should be taken by the working microscopist to protect his eyes from the direct light and heat of the lamp ; and work should be discontinued, as soon as marked visual fatigue is experienced.—S. Y. H.

thickness,—say 0.15 mm.,—and may be either square, oblong, or circular in form. For objectives of very high power, cover-glasses of special thinness are requisite.

Other articles of glassware used are: watch-glasses; glass rods; tubes of various sizes, capillary, etc.; several flat-bottomed, shallow vessels, or saucers; bell-glasses; beaker-glasses; wash-bottles; bottles for reagents, etc.; retort stand, with retorts, funnels, etc.; a graduate.

A piece of black glass, and one of white porcelain, of convenient shapes, will be found very serviceable as beds, when mounting specimens. White, or unstained objects are best seen over the black surface; while the white porcelain is always used in manipulating colored preparations.

3.—Metal Instruments.

As for needles, forceps, scissors, knives, spatulas, etc., the necessity of maintaining these in an absolutely fault- less condition cannot be too strongly insisted upon; and the example of many older microscopists, who work with blunt, rusty needles, with scissors which do not cut, for- ceps which do not hold, etc., should be carefully avoided. Our instruments must be kept as bright and sharp as those of an oculist.

The *razor* should be ground flat on the lower side, and be kept clean and sharp. In spite of double-knife and microtome, the razor is still often needed in the preliminary examination of tissues ; and, in using it, the whole length of the blade should be utilized, while traction, rather than pressure, should be resorted to. After a few weeks of practice, sufficient skill is acquired by one to enable him quickly and surely to make uniformly thin sections, both of fresh and hardened specimens ; and, in many cases where absolutely exact or extremely thin sections are not required, this method is preferable, because of its simplicity.

The *double-knife* is much used also, and especially in the investigation of fresh specimens. With it, one obtains large, uniform sections, even from fresh and soft organs, though at the expense of much material ; as the tissues are frequently hacked in a most unhandsome manner. The double-knife is drawn through a specimen with a quick stroke ; and, in order to avoid injuring the edges of the blades, a soft support should be provided, such as a piece of fresh liver. By most instrument-makers, the spring between the two branches of the double-knife is made too stiff ; and I have often found it advantageous to remove this altogether.

To set the blades, they are screwed closely together, the screw which is nearer the cutting edges being first completely loosened to permit this ; after which, the latter is turned in a reverse direction, till the knife-edges are torced slightly apart. The blades should be as nearly parallel as possible.

4.—THE MICROTOME.

No microscopist of the present day would be willing to dispense with this instrument. Possessed ten years ago by only a few, microtomes are now widely distributed; and it is certain, that much of the progress, made in microscopical technology, may be attributed to this circumstance.

It would be extremely tiresome, and, indeed, it is unnecessary to describe here the peculiarities of the different microtomes, especially as new instruments, or modifications of old ones, are being devised every year. We can employ almost any style ; though I would advise against such models as require the imbedding of the specimen, from which sections are to be made, in paraffine, etc.

In Europe, good microtomes are made by Dr. Long, of Breslau ; Katsch, an instrument-maker of Munich ; and by the mechanicians Schanze, of Leipsic (Patholog. Institute) ; Jung, of Heidelberg ; Meier, of Strasburg ; and various others. The author worked for a long period with Long's instrument, but is so pleased with Schanze's new model that he would especially commend it.*

---

* Schanze informed me that he would pack, ready for shipment, his instrument, with two knives and the freezing attachment, for 103 marks

The principle of the instrument is simple. The knife
is clamped to a carriage, which moves back and forth in
a groove ; while the specimen, to be cut, is elevated by
turning a horizontally disposed wheel, which is graduated
at its periphery.

When the knife is advanced, a section is made of a
definite thickness, corresponding to the arc through
which the adjusting wheel is turned. In Long's microtome, the specimen is raised by advancing its holder up
an inclined, graduated plane ; while in Jung's instrument,
this is accomplished with great accuracy by means of a
fine micrometer screw. In the latter construction, the
movement of the carrier is also remarkably smooth.

The object to be cut is fixed in a clamp, of which several,
varying in size and form and easily interchanged, may be
employed In any case, it is of importance that the specimen be firmly fixed ; and, to that end, it is customary to
clamp it between two flat pieces of well-hardened liver
(amyloid liver, hardened in alcohol, is best). The hardened specimen—a firm, uniform consistence is a necessary
condition in the preparation of good microtome sections
—is put between two pieces of liver, and so fixed in

(about $26.00). Unfortunately, the import duty of 45 per cent. on
instruments seriously enhances the cost for Americans. An efficient
substitute however, though without the freezing apparatus, is M.
Rivet's pattern, which is furnished by our dealers for $15.00.—S.
Y. H.

the clamp, that its upper surface projects 1-2 mm. above
the metal. The liver acts as a fixation splint, holding
that portion of the specimen, which projects above the
clamp, sufficiently firm to enable it to meet the advancing
blade without yielding ; and thus perfect and complete
sections are secured. Or, again, a cork may be fixed in
the clamp, to the upper surface of which a portion of the
specimen several mm. in thickness has been cemented,—
using for this purpose a thick solution of gum arabic, or
the so-called fluid gelatine, both of which harden quickly
in alcohol (see " Imbedding "). In this way, very little of
the preparation, to be examined, is used ; and, inasmuch
as the cork alone is fastened in the clamp, it is subjected
to no pressure. This procedure, which, to the best of my
belief, originated with Weigert, is especially expeditious,
and deserves great commendation.

The object being properly clamped, the knife is so
arranged that the whole blade is used, from heel to point,
—a matter which is very often neglected. The beginner
frequently forces the blunt end of the blade against the
specimen, or, on the other hand, sets the knife so that
only half of its cutting edge is utilized.

Both of these extremes should be avoided. Fixed
so that the distance between two lines, drawn through
the ends of the cutting-edge and parallel to the line of
motion, shall slightly exceed the width of the object

which is to be cut, the blade meets the specimen at the most favorable angle, and cuts rather than pushes its way through it.   The narrower the object, the less the obliquity of the blade, and the more perfect the section. When cutting, the object and knife should be kept well moistened with alcohol,* and the working parts well lubricated, preferably with neat's-foot oil.

Instead of the clamp for hardened objects, a freezing-plate, or box, may be attached to the microtome, against the under surface of which an ether-spray impinges, causing, by its evaporation, an intense cold.   If a piece of fresh tissue be laid upon the freezing-plate, it is frozen almost immediately ; and one is thus enabled to prepare very fine and uniform sections, without any previous hardening.

Thick and Thin Sections.

One can vary the thickness of the sections at pleasure, by raising the specimen to a greater or less degree, reading off the same in hundredths of a millimetre from the scale, which is usually attached to microtomes.

---

* This is best accomplished by attaching a rubber tube, provided with a mouth-piece, to a small-sized wash-bottle, so disposed, that, when air is forced into it, a stream of alcohol is projected upon the knife-blade as it passes ; whence it flows to the object, as soon as the two come in contact.   This method of procedure leaves both hands free, and does away with the frequent interruptions, otherwise unavoidable.—S. Y. H.

The beginner, as a rule, believes in the necessity of obtaining the thinnest possible sections in all cases ; while the more experienced observer often deals intentionally with those of considerable thickness. Against sections of extreme thinness may be urged the following objections :

In the first place, they are manipulated with difficulty; and considerable time is often lost in spreading them upon the slide.

Secondly, the various elements, contained in the meshes of thin sections, are very apt to fall out ; and, as these are generally of extreme importance, the object of the examination may be defeated. For this reason, also, the pencillings and agitations, to which thick sections of normal tissues are frequently subjected, in order that their transparency may be enhanced, are to be commended only with certain restrictions in the investigations of pathological anatomy.

Thirdly, structures which are sparingly distributed throughout an organ, as, for example, animal and vegetable parasites, are naturally more apt to be discovered in thick sections, provided the latter are sufficiently transparent, and the objects well contrasted with the investing structures.

Fourthly, in thick sections definite stereometric conceptions of the structure of an object are frequently obtained, inasmuch as several superimposed strata are scanned directly, *in situ et in continuo ;* while with extremely thin sections, plane images alone appear. On the other hand, it is clear that many structures of particular delicacy are visible only in very thin sections ; as, in the thicker ones, they are completely masked by the many superimposed contours.

For most purposes, sections of fresh organs, ranging in thickness from 0.05—0.10 mm. (0.002—0.004 inch), answer very well ; though, in the case of hardened preparations, thinner sections, ranging from about 0.01— 0.03 mm. (0.0004—0.0012 inch), are generally employed, particularly when stained.

A very essential advantage of the microtome lies in the fact, that sections of any desirable thickness and number may be prepared with great ease and certainty. And, indeed, one usually cuts many more sections than are required at the time being, the remainder being set aside in a small vial filled with alcohol, to be used in the future, should occasion arise.

Further Treatment of the Sections.

From the blade of the microtome, double-knife, or razor, the sections are transferred to a watch-glass, filled with some fluid, and, for this purpose, a soft brush, previously moistened, answers best.   For fresh or frozen

sections, this fluid is a solution of common salt, in which
they are also examined ; while sections of preparations
hardened in alcohol, are first placed in alcohol, then
usually in distilled water. During the manipulations,
which the sections subsequently undergo, they may
easily be rendered unfit for use. They must be lifted
up and transferred to the various vessels, which contain
the necessary reagents and staining solutions, to be
finally placed uninjured upon the slides. For this purpose, thin spatulas of copper, platinum, or nickel-plated
steel, turned up at the end to an angle of about 35°,
answer best ; and these are carefully passed under the
sections, as they float in the watch-glass. In this way
alone, can a fine section be raised out of one fluid, and
deposited into another without folding. The centre of
the object-glass, or slide, must also be covered with
a deep stratum of fluid before the section is transferred
to it ; and, then, no force should be employed, the fragment, or section, being allowed to glide down from the
spatula to the slide with its investing fluid. Now, the
cover-glass is applied ; and the superfluous fluid is removed, either with a capillary tube, or by means of
blotting-paper. During these procedures the slide lies
on a plate of black glass, when the object is not colored;
otherwise, upon one of white glass,—both hands of the
microscopist being free. Holding the slide in one hand,

while the object is being placed in position with the other, is the awkward practice of many beginners. In order to secure the delicacy of touch, requisite in these and other manipulations, I would advise that the forearm and ulnar border of the wrist be rested firmly upon the table ; as it is very difficult to mount fine sections well, when the arm is unsupported.

Sections of tissues, hardened in alcohol, are removed from the watch-glass of alcohol into one filled with distilled water ; where, amid the lively movements due to the diffusion of the two fluids, they flatten out, and change from a shrunken, wrinkled condition, into transparent laminæ. For examination, they are spread out upon the object-glass in distilled water, which, in the majority of cases, is then displaced by glycerine, from the edge of the cover-glass. Large and thin sections are spread out in viscid glycerine less easily than in distilled water ; while smaller sections can be brought directly into a drop of glycerine.

III. REAGENTS.

MICRO-CHEMISTRY.

Reagents should be kept in glass bottles, fitted with stoppers of ground glass. While this rule is observed in every chemical laboratory, and, indeed, in every apothecary's shop, one continually sees microscopists employing bottles, fitted with unclean corks—a practice which is very objectionable. For those reagents which are in constant use, double precautions may fittingly be taken ; and the author is accustomed to keep his glycerine, acetic acid, distilled water, oil of cloves, Canada-balsam solution, etc., in bottles whose stoppers have rod-like prolongations reaching nearly to their bottoms, while the whole is covered with a tightly-fitting glass cap. By this cleanly and careful management of his reagents, the beginner is accustomed to the most painstaking neatness, while working.

As a matter of course, we always employ *chemically pure* preparations ; though, as yet, some staining agents, which are in use, do not possess a fixed chemical value.

In addition to the above-mentioned precautions, taken on the score of cleanliness, it is necessary to make use of the reagents always with an eye to the maintenance of their purity. The microscopist, who obtains a drop of any reagent by dipping a needle, brush, or even his finger into it, will never be able to make investigations of the finer sort, particularly where bacteria are concerned. Only a carefully cleansed glass rod, or a glass tube freshly raised to a red heat, should ever be brought into contact with a reagent.

"*Artificial Products.*"

The employment of reagents in histological investiga-tions is of the highest importance, for many structural elements can be studied only with the aid of chemical influences ; but we must naturally endeavor to examine these elements in the most unchanged, natural, and, when possible, in a living condition.   And yet, he would be going very far astray indeed, who should positively reject, as "factitious products," all those structures, which are rendered visible only by the action of certain reagents.   Ordinarily, there are no signs of a nucleus in the living white blood-corpuscle, or of cells in the living cornea ; because the differences in the refractive power of nucleus and protoplasm, cell and ground-substance, are too small to insure detection ; or, again, because the investing substance is not sufficiently transparent to dis-close the delicate outlines of the inclosed body. When death supervenes, however, these distinctions grow more apparent, by virtue of certain chemical changes, as coagulation, etc., or the investing material be-comes more transparent ; and we are surely justified in assuming the presence of nucleus and cells, in the living white blood-globules and in the living, cornea, respectively, though we are able to demon-strate them only after the occurrence of post-mortem changes.   This is true also of the limiting membranes of many epithelial cells, of the axis-cylinders of nerve fibres, etc., which are really present, though invisible during life. Nevertheless, in the investigations of pathological anatomy, we see the tissues almost never in a living, natural state, but always in a condition of more or less advanced cadaveric change. Accordingly, we must always bear in mind the fact, that a structure of constant occurrence *post mortem* was not of necessity present, as such, *intra vitam ;* but, at all events, such a phenomenon suggests that a distinction existed even during life, which has only been rendered visible by post-mortem change, or by the particular reagent employed by the observer. *Furthermore, if among a certain number of elements, originally identical so far as appearances are concerned, some react in a peculiar manner when a particular reagent is used, while others do not,—when, for example, some are colored by a certain staining agent, while others remain colorless, we are necessarily forced to conclude that a real difference preëxisted.* Upon this principle are based all the various and oftentimes complicated preparatory methods, which we employ for the demonstration of the different histological elements. From this simple consideration it becomes evident how we ought to regard the microscopical effects of our reagents. We should not permit ourselves to be staggered by the scornful designation, " art products," remembering rather, that in times past many histological discoveries of great moment were thus discredited at first. *In our microscopical investigations we are, for the most part, engaged, not as observers merely, but we experiment; and our results are accordingly compounded of preformed structures, on the one hand, and, on the other, of factors introduced by ourselves.* To pass judgment, then, upon the nature of an object at first sight, and without reference to these aids,—for the most part so simple,—would be to expose one's self to the greatest errors. A seeming fibre, for example, can represent an actual preformed fibre, a crease, a product of coagulation, etc. In our pathological investigations particularly, we make another use of our reagents. We are often enough called upon to search for particular elements, foreign bodies, parasites, etc. ; and, if we know that the particular objects of our quest resist the action of certain reagents and modes of treatment by which the remaining substances are destroyed, we are possessed of a method of examination most serviceable for our definite purposes, even though the structural integrity of the specimen be completely sacrificed thereby. From this it appears, that the use of the microscope can-not always be regarded as something purely mechanical, especially when questions in pathology are concerned; but rather that it frequently demands the exercise of a certain amount of reflection and circumspection, in choosing the course to be pursued.

Micro-Chemical Examinations.

Micro-chemical investigations are conducted by exposing the object to be examined to the action of reagents for a variable length of time, either by placing it in a watch-glass containing the particular agent to be employed, and afterwards transferring it to a slide for examination, or by effecting this action while the object is under the microscope. In the latter case, a drop of the reagent is placed upon the slide at the edge of the cover-glass, whence it gradually flows toward the specimen,—a process which may easily be expedited by slightly moistening the edge of a slip of blotting-paper, and bringing it gently into contact with the opposite edge of the cover-glass, thus creating a current toward the absorbent.

When this method is employed, one can readily follow the various effects of the reagent, such as the solution of protoplasmic granules, red blood-corpuscles, and lime salts under the action of acids, etc.

In doing this, the beginner must take special care that the reagent be not deposited *upon* the cover-glass ; as the objective may easily be injured by contact with a corrosive fluid. Other more complicated reactions, and particularly the majority of staining processes, are carried on in watch-glasses,—the precise effects being learned by comparing the resulting preparation with "control-specimens," or the object is sketched both before and after the treatment. The most generally employed re-agents are the following :

I.—DISTILLED WATER.

Small amounts of foreign material are generally held in solution by distilled water, and, in summer particularly, this agent affords ample sustenance for various micro-organisms,—a fact which should be remembered, when such organic structures appear in preparations which have been treated with it. This difficulty may easily be overcome by frequent boilings.

Distilled water is very rapidly diffused throughout most of the constituents of fresh tissues, and these are thereby more or less extensively altered. At all events, the vital properties of isolated elements of the human body cease very soon in this fluid ; while the dead cellular elements of the cadaver undergo considerable change. This is most strikingly illustrated in the case of the red blood-corpuscles, which swell, yield up their coloring matter, and soon become completely invisible in water. From this, the use and limitations of distilled water, when employed in the investigation of fresh tissues, become at once apparent. We employ it with predilection when, in the examination of sanguineous tissues, it becomes necessary to quickly remove from view the blood globules, which are frequently present in such numbers as seriously to interfere with the view of the remaining structures. At the same time we are never to forget that the tissue itself can be essentially altered by the action of distilled water.

In the case of tissues hardened in alcohol, distilled water simply causes a swelling, and, for the most part, this is uniform in character, so that the original proportions are nearly restored. Further changes are but seldom produced ; as the principal constituents of the tissues—the albuminates—have become coagulated,—that is, they have undergone a change which renders them insoluble in distilled water. But, of course, it must be borne in mind that those substances which are soluble or diffusible in water, as, for example, glycogen or sugar, are quickly extracted from the sections.

2.—Chloride-of-Sodium Solution.

Indifferent Media.

Owing to the fact that the most various micro-organisms are generated quickly and in great numbers in these fluids, they should be renewed frequently. The addition of antimycotic substances is strongly advised against for the sake of reliable results ; but the chloride-of-sodium solution can be easily sterilized by boiling.

To preserve the protoplasm and the red blood-cor-

puscles as intact as possible, an 0.8-per-cent. solution
of sodium chloride is used, which medium is the one
ordinarily employed in pathologico-anatomical investigations for sections of fresh tissues, as well as for the
dilution of fluids.   If it be specially desirable to maintain for some time the life of the cells, in order that their
properties may be investigated, one part of the white of
egg is added to nine parts of the salt solution (so-called
artificial serum) ; or, aqueous humor, fluid from hydro-
celes, serous exudates, blood serum, amniotic liquor, etc.,
are employed.

3.—ABSOLUTE ALCOHOL.

(*Hardening.*)

We always make use of the purest alcohol, never of
the diluted variety or proof spirit ; as the latter, in addition to its admixture with water, is always contaminated
with other substances, and often exhibits an acid reaction.   In case we desire to employ diluted alcohol, however, we mix the absolute variety with the requisite
quantity of distilled water.

Alcohol, diluted with two parts of water, is sometimes
employed as an aid in isolating tissue elements (Ranvier).

In it the cells become more capable of resistance,
while the connecting structures remain soft ; so that the
isolation of cells, which, before, seemed intimately united

with each other and with the intercellular substance,
becomes an easy matter. To effect this the fresh tissues
are placed in the thirty-three-per-cent. alcohol for about
twenty-four hours.*

Alcohol finds its chief employment, however, in hardening tissues to a consistency suitable for section-cutting. This hardening effect of alcohol is due, essentially,
to two moments, viz.: the removal of water and the
coagulation of albuminates ; although alcohol also
removes from the tissues some extractive matters, which,
morphologically considered, are unimportant, besides a
small amount of fat. From this latter effect, it is very
evident, that, to obtain a correct estimate of the amount
of fat contained in certain specimens,—as in tissues
which have undergone fatty degeneration,—these should
always be examined in a fresh state, never after previous
hardening in alcohol. By reason of the removal of
water, the tissues undergo diminution in size ; and,
except the water be uniformly distributed throughout the
specimen, the amount of shrinkage varies, and an un-
wished-for distortion ensues. For the most part, however, the original form is quite well preserved ; and
inasmuch as the sections, when placed in distilled water,

---

* Frey explains this action of very dilute alcohol differently, claim-
ing that it acts as a macerating medium. This apparent anomaly is
confirmed by the analogous action of weak solutions of chromic acid.
—S. Y. H.

imbibe an amount about equal to that originally possessed, they become very like corresponding sections in
the fresh state. The principal point of distinction is the
loss of transparency, which is due to the granular, coagulated albuminates, and persists even after the sections
are immersed in water. This drawback is practically
removed by the use of glycerine as a clearing agent, or
of acids and alkalies, by which the precipitated albuminates are again dissolved, though only at the expense of
many other structures.

To harden a specimen, it is best to divide it into small
pieces, and to place these in a large volume of absolute
alcohol, taking care that each piece is completely invested
by the fluid. In this way a piece of tissue comprising
2–3 c.c. may be thoroughly hardened in twenty-four
hours,—smaller portions in a much shorter time. The
method, formerly much in vogue, of laying the tissues
first in weak, and then gradually in stronger solutions of
alcohol, has justly been abandoned.

Alcohol is the agent best adapted to the hardening of
most tissues, and is employed by us almost exclusively
for this purpose.*

It is of special consequence in the case of pathologico-
anatomical objects, that the changes wrought by the

* It has already been explained, why hardening in alcohol is not
adapted to tissues, which have undergone fatty degeneration.

hardening process should be simple and easily controlled. These conditions are fulfilled, when alcohol is the agent used ; while hardening with the salts of chromic acid, formerly such a popular method, causes many altera- tions—cloudings, discolorations, etc.,—which vary with the amount of time consumed, the temperature, etc., and are extremely difficult to control.

Some little expedients, which become necessary at times, almost suggest themselves. Thus, for example, the globe of the eye is wont to shrink very quickly in alcohol, an evil which is easily remedied by injecting alcohol into the vitreous humor with a hypodermic syringe until the rotundity of the organ is restored. This is done early, and may have to be repeated.

Many tissues, particularly the lungs, muscles, etc., do not become sufficiently hardened for section-cutting, even after a long immersion in alcohol ; and it is advisable to lay such preparations for twenty-four hours in a diluted mucilage (mucilaginis acaciæ et glycerini, partes æ- quales). If, now, the specimen, saturated as it is with gum arabic, be again placed in alcohol, it hardens very uniformly and completely, inasmuch as the acacia is pre- cipitated by the alcohol. The gum is soon removed from sections of such preparations by the action of water.

Only for the *central nervous system* is alcohol poorly

adapted, and, least of all, for the white substance ; as it fails to produce a sufficient degree of hardness,—the defect being proportional to the diminished amount of water present in nervous tissue. Moreover, a large share of the fatty constituents of the medullary substance are extracted by the alcohol, to be precipitated, later, in a crystalline form, thereby injuring the tissue considerably. For these important structures, therefore, we find it impossible to dispense with the chrome salts.

4.—ETHER AND CHLOROFORM.

(Removal of Fatty Substances.)

Both of these agents are frequently employed for the removal of fat, though not from fresh tissues, naturally ; as these are saturated with water, a fluid for which ether and chloroform possess only a very slight affinity. Accordingly, the structures—sections, for example—must first be deprived of their water by the action of alcohol. Thus, a section from which the fat is to be removed is first placed for about five minutes in a watch-glass, or other shallow receptacle, filled with absolute alcohol, and thence into a similar vessel, filled with ether or chloroform. Should the latter fluids become cloudy, the presence of water is indicated, and a second bath in absolute alcohol becomes necessary. After remaining in

the ether or chloroform a few minutes,* or until the
soluble substances are completely removed, the section is
again placed in alcohol for a time, and subsequently in
water. We examine the section in pure water ; though
the marked opacity, produced by the coagulated albuminates, usually requires the addition of acetic acid, which
acts as a solvent. The order to be observed, then, in removing the fat from fresh sections is as follows :

Chloride-of-sodium solution, or simple water ; alcohol ;
chloroform or ether ; alcohol ; water, to which acetic
acid is usually added.

5.—ACIDS.

a. Sulphuric, Hydrochloric, and Nitric.

(*Decalcification.*)

The strong mineral acids, in the higher concentrations,
possess the property of quickly coagulating albuminous
bodies, and can in consequence be used with advantage
in the fixation of certain very delicate structures. Thus,
for example, in the case of the so-called " nuclear
figures " (Kernfiguren) of Altmann,† it is best to lay the
small pieces of tissue in a 3-per-cent. solution of nitric acid

---

* Sections immediately become very transparent in chloroform,—
not, perhaps, because of the solution of the fat, but, rather, by reason
of the high refractive power of chloroform. When returned to alcohol,
the section exhibits its former opacity.

† Altmann : *Arch. f. Anat. und Physiol.*, 1881, S. 219.

(sp. gr. 1,020) for a short time,—say an hour,—after which they are washed in distilled water, and hardened in absolute alcohol.   Flemming and other authors employ a still stronger nitric-acid solution for this purpose.

In a very dilute state (1 to 1,000), the mineral acids produce, essentially, a tumefaction of the majority of protoplasmic substances, of the contractile and gelatinous tissues, etc., resembling acetic acid in this respect.   Again, these are employed for the removal of calcareous material.   In order to obtain good sections of calcified structures, such as bones, teeth, tumors which have undergone calcareous change, etc., the lime salts must be removed,—a desideratum soonest satisfied by the use of hydrochloric or nitric acid, the older method of grinding such material to a proper degree of thinness being seldom resorted to.

A dilute acid is used ; and this is mixed with alcohol containing chloride of sodium (Von Ebner), in order to avoid the swelling of the ground substance.   The mixture contains the following ingredients :

Hydrochloric acid . . .	5.00
Alcohol . .	1,000.00
Distilled water	200.00
Chloride of sodium	5.00

This decalcifying fluid must be changed often ; when it affords really excellent results.   A solution of chromic acid (about 1 per cent.), or a saturated solution of picric acid (Ranvier), acts somewhat more slowly.

When examining a specimen with the microscope, one may come across dark deposits, which he is disposed to regard as lime salts ; and that such is indeed the fact, he feels convinced, if the dark contour disappears, when exposed to the action of a mineral acid. In most cases the lime is in combination with carbonic acid ; and, if an acid be added, a lively evolution of gas ensues, affording a very pretty and striking picture under the microscope. If we use sulphuric acid, the sulphate of calcium, or gypsum, is formed—a slightly soluble substance, which soon crystallizes in beautiful prismatic columns, and these are frequently grouped together in tufts. The gypsum crystals are exquisitely doubly refracting, a fact which is appreciated, when these are viewed through a Nicol's prism, placed above the eye-piece.

Many connective substances, tissue cements, are dissolved in strong acid solutions. Thus a 20-per-cent. solution of nitric acid, for example, is employed in isolating the smooth muscular fibres, as is also a 33-per-cent. solution of caustic potash.

To isolate the uriniferous tubules for long distances, that their complicated course may be established, strong hydrochloric acid is used, aided, sometimes, by an elevated temperature. This method, however, has not yet been adopted in the study of nephritis.

b. Acetic Acid.

The employment of the organic acids, and of acetic acid in particular, is very general among our microscopists. We use them, principally, to effect a solution or tumefaction of the albuminates, and of the gelatinous substance, of which, as is known, the connective-tissue fibrils consist.

Inasmuch as the nuclei of cells, elastic tissue, fat, nerve medulla, etc., resist the action of acetic acid, it proves a very convenient agent for exposing to view the concealed nucleus of a dark granular cell, as well as the elastic tissue, which is distributed throughout connective and muscular structures. So, too, the fat-granules, present in protoplasm, in the contractile substance of muscles, etc., become much more evident after the action of this agent ; as do also any *micro-organisms*, present in the tissues. And so marked is this specific action of acetic acid, that a 1-per-cent. solution is effective ; while, even in the proportion of 1 to 1,000, the clearing effect becomes very apparent, though somewhat more slowly.

When a section from a fresh specimen, or from an alcoholic preparation, is placed in acetic acid, it usually becomes quite transparent, and, at the same time, increases markedly in size. For the most part, this tumefaction is not uniform, however ; and the section assumes a coarsely undulatory appearance, which wellnigh unfits it for examination. It is better, therefore, to apply the acid, when the section is under the cover-glass,—a slight elevation of the latter sufficing to admit the drop of acid which has been deposited at its edge ; while the slight weight of the cover suffices to preserve the tissue flatly extended. Should a prompt and decided action be desirable, we employ the undiluted acid— *acidum aceticum glaciale*— ; though; for most purposes, it is advisable to dilute this with a little distilled water.

In many substances, saturated with alkaline albuminous solutions, an opacity at first ensues on the addition of acetic acid, caused by the neutralization of the alkali ; but, if more acid be added, this is dispelled, and the object becomes clear. And yet, a lasting opacity can be produced by acetic acid,—one which is not resolved even by an excess of the acid,—when this agent is added to tissues or fluids containing *mucin*, this substance being precipitated. Fibrin, serum-albumen, and mucin are often found together in exudates, the contents of cysts, etc. ; and the action of acetic acid will vary according to the quantitive proportions of the mixture. In most cases, substances are markedly cleared, and rendered transparent, by the action of this acid.

In view of the pronounced increase in volume, produced in albuminous and gelatinous bodies by acetic acid, it is by no means remarkable, that the contours, which appear after the action of this acid, *i. e.* the boundaries of those structures which resist the action of the agent, do not always remain quite unaltered : indeed, as early as the fifth decade of the present century, Henle called attention to the fact, that the multiple nuclei appearing in white blood-globules, pus corpuscles, etc., when acted upon by acetic acid, are not preformed structures ; but rather, that these cells contain but one nucleus, which is broken up into several fragments by the action of the acid.

In many instances this is no doubt true ; and, although we now know that many lymphoid cells possess several nuclei during life, we are earnestly advised to remember, that the nuclei, which appear in a cell after it has been subjected to the action of acetic acid, may possibly be factitious products—fragments of what was originally a single nucleus. In structures which contain connective tissue, considerable distortion is often caused by acetic acid. For example, the regular longitudinal rows in which the cells of tendons, fasciæ, etc., are disposed, cannot usually be demonstrated after the tumefying action of the acid, the nuclei then appearing irregularly arranged and scattered. Ranvier took the precaution to fasten the ends of a small tendon, which was stretched upon a slide, using for this purpose little balls of wax. After the cover was in place, he allowed acetic acid to act slowly upon the object ; when he found, that the tissues swelled more uniformly, and the longitudinal arrangement of the cells became evident. The tyro is earnestly advised to convince himself of the truth of these, and similar, facts by his own experiments ; in order that he may be able to judge what structural changes he produces with his various reagents.

Formic and tartaric acids are less used, their effects being similar to those of acetic acid.

c. Picric Acid.

Picric acid, on the other hand, fulfils a twofold office, being both a hardening and a coloring agent. In a saturated solution of this acid, the albuminates are gradually changed into the insoluble modification ; so that the tissues attain a consistence suitable for section-cutting, almost entirely without shrinkage. At the same time, the majority of substances are colored yellow, some more deeply than others ; as, for example, the smooth muscular fibres, the cornified cells of stratified epithelium, of the epidermis, etc.

Sections made from specimens hardened in alcohol, also take this characteristic color very nicely and quickly, only a few minutes being required ; but the color is rapidly removed by water or alcohol. Should we desire to preserve the stain, a small quantity of picric acid must be added to the water, alcohol, glycerine, or other fluid, which is employed.

d. Chromic Acid.

Chrome Salts, Müller's Fluid.

In a very dilute state—say 1–10,000 or 1–20,000—chromic acid is employed as a macerating agent. If a small section of the spinal cord, for example, be allowed to remain for twenty-four hours in such a solution, it becomes an easy matter to isolate the ganglion cells with their branching prolongations ; inasmuch as the intercellular connective substance is softened, or dissolved.

For our purposes, however, more importance attaches to the acid and its salts in the rôle of hardening agents, the former being employed in a 0.2- to 1.0-per-cent. solution. Of the salts, the bichromates of potassium, or ammonium, in about a two-per-cent. solution, are most used. Müller added to the solution of the bichromate of potassium some sulphate of sodium, in the following proportions :

Potassium bichromate .	2.0
Sodium sulphate	1.0
Distilled water	100.0

and this *Müller's Eye-fluid* is much used for hardening nervous tissues, as well as the eye, though originally recommended for the retina.

Weeks and months elapse, however, ere the requisite hardening ensues, the duration of the process being proportional to the size of the immersed specimen ; as the chrome salts penetrate the tissues very slowly. For a cerebral hemisphere, from six months to a year must be allowed, the fluid being changed often.*

To prevent moulding, a small piece of camphor may be added to the solution (Klebs). According to Weigert, the hardening takes place much more rapidly in an incubator at a temperature of 30°–40° C. (86°–104° F.).

Erlitzki suggests the following hardening fluid :

Bichromate of potash . .	2.5
Sulphate of copper	0.5
Distilled water	100.0

In this, the hardening requires only 8–10 days, at an ordinary temperature. Subsequently, the preparations are placed in alcohol, which may be somewhat diluted. Portions of the central nervous system, when thus exposed to the action of chromic acid or its salts, become thoroughly and uniformly hardened ; while, at the same time, certain characteristic chromatic changes take place, which are apparent even to the naked eye. The gray

---

* This time may be shortened to five or six weeks by incising the organ in various directions ; so that the fluid can come in direct contact with every cubic inch, while the relations of the parts are not sacrificed. A thin-bladed, keen knife should be employed, thereby reducing the necessary cutting-pressure to a minimum.—S. Y. H.

substance becomes lighter than the white,—the latter assuming a dark green shade.

The ordinary form of gray degeneration, or sclerosis of the white columns of the cord, is distinguished by a dark brown tone ; while most secondary degenerations are marked by a lighter color than that assumed by the normal columns ; and this, too, in cases which disclosed absolutely no distinction, in color, between normal and degenerated parts, while in a fresh state.

For other organs, the hardening in solutions of chromic acid or its salts,—formerly so much in vogue,—is to be recommended only in a few exceptional instances ; and, save for the nervous tissue and the globe of the eye, we greatly prefer alcohol, with the occasional aid of the glycerine-mucilage mixture already mentioned. Filiform or retiform coagulations are often produced by the salts of chromic acid, which may be erroneously regarded as preëxisting structures ; while the dark, granular precipitates, produced by them in cells and intercellular substance, aɪe oftentimes very annoying, and difficult to clear up by chemical action. Calcareous deposits are slowly dissolved by chromic acid and the bichromates, and can thus be removed from the field.

As a rule, micro-chemical reactions fail, in the case of specimens prepared with the chrome salts ; and, for this and other reasons, we recommend them as hardening agents only where alcohol proves unsatisfactory on account of the peculiar chemical nature of the object, which is essentially true only of nervous tissue, or of organs which have undergone marked fatty degeneration. At the same time, however, it is always desirable to compare the results with corresponding sections from fresh, or alcoholic preparations ; inasmuch as any calcareous matter, which may chance to be present, can easily be dissolved by the chromic acid and its acid salts, and thus escape observation.

The mono-chromate of ammonia, in a 5-per-cent. solution, was employed by Heidenhain with great advantage in the study of the kidneys, and, particularly, for the demonstration of the bacillar structure of the epithelial cells in the uriniferous tubules. The agent might also be recommended for pathologico-anatomical investigations.

6.—ALKALIES.

Solutions of Soda, Potassa, and Ammonia.

Alkalies cause a resolution or swelling of the albuminates, gelatinous structures, the contractile substance of smooth and striated muscles, and of the *nuclei of cells ;* even the corneous tissues are rendered thoroughly transparent by their action.

Resisting the action of the alkalies, we have :

1. Elastic tissue.

2. Fats, including the medulla of nerves.

3. Lime, pigment, etc.

4. Amyloid material ; as well as chitine,* cellulose, mycelium, spores, and bacteria.

From this, the importance of the rôle, assigned to the alkalies in our work, becomes manifest ; they come into use, whenever any of the objects in the latter category are to be looked for.   It is true, that, by their use, the structural qualities of a tissue are almost completely sacrificed ; and, while the resisting nuclei of a section, treated with acetic acid, give us our bearings tolerably well, such is not the case with the alkalies, sparing, as they do, only the elastic tissue and homogeneous membranes for our guidance.

For most purposes, a 1- to 3-per-cent. solution of caustic potash or caustic soda is employed ; and, even in this dilution, the clearing effects begin immediately.   The use of a more concentrated solution—33 per-cent.—is attended with a peculiar effect : most elements are preserved, while the connective substance is dissolved. This is particularly true of smooth and striped muscular fibres.   If a small portion of a uterine myoma be placed in a watch-glass, filled with a 33-per-cent.

---

* The horny substance in the tegumentary system and other parts of the crustacea, arachnida, and insects, including the hooklets of the tæniæ.—S. Y. H.

potassa solution, for a few minutes, it will break down under the needle almost of itself, and be resolved into its individual fibre cells. Care should be taken only, that the solution be not further diluted ; since, in that case, the fibres themselves are dissolved. The preparation should therefore be examined, also, in the 33-per-cent. fluid. Red blood-corpuscles, as well, preserve their form in a solution of this strength, while in the higher dilutions they disappear immediately.

According to an observation of Virchow's, weak solutions of the alkalies are able to re-excite the movements of ciliated cells, which before were motionless, and, to all appearances, dead.

7.—GLYCERINE.*

Glycerine has an especial value in the histological examination of tissues which have been hardened in alcohol, and other agents which coagulate albumen, such as picric and chromic acids, and their salts ; for, by their use, marked opacities have of necessity been produced in the sections ; and, if we employ acids or alkalies to dissolve these albuminous granules, thus precipitated by the hardening agent, we destroy at the same time many

---

* The glycerine used must, above all things, be free from an acid reaction ; a slight admixture with water is of less consequence. We usually employ it in a pure state ; as, when diluted with water, it is prone to mould.

other structures, as the connective-tissue fibres, fibrine, and blood corpuscles. In such cases, therefore, we use glycerine as a clearing agent, which effects the wished-for result,—not by a chemical solution or swelling of the albuminous granules (fat alone being gradually dissolved in glycerine), but rather by virtue of a purely physical quality ; viz. its great refractive power. That glycerine really possesses this property, we can immediately convince ourselves by comparing the contours of glass rods, dipped in glycerine and water, the outlines in the former case being far more delicate, or less distinct. Or, if we saturate a piece of filter-paper with water, and another with glycerine, the latter becomes much more transparent than the former. So, too, the contours of the elementary constituents of tissues, which are saturated with glycerine, are all rendered less distinct ; whence this agent is wellnigh useless in the investigation of fresh tissues, whose elements already possess very slightly marked outlines, since these outlines are thereby rendered almost completely invisible. On the other hand, the clearing effect of glycerine is nicely adapted to preparations hardened in alcohol ; and, indeed, it may be said, that the examination of tissues, so prepared, has been developed to its full fruitfulness only since the introduction of this agent into the laboratory of the microscopist.

Glycerine mixes with water, as well as with alcohol,

acetic acid, etc., in all proportions, though quite slowly, it is true, in accordance with its syrupy consistence. If, therefore, we desire to produce a chemical reaction very quickly, as, for instance, to determine the effect of iodine or of an acid, a glycerine preparation is less suitable. At all events, the removal of the glycerine from a specimen is an easy matter, it being only necessary to place the same in water. Moreover, as we know, glycerine neither evaporates, nor undergoes other chemical change, when exposed to the air ; but, at the most, absorbs some water, when the atmosphere is charged with moisture. This quality makes glycerine a superior preservative agent for microscopical sections ; and, should it be desirable to preserve a specimen which is in water or a watery solution, it is only necessary to place a drop of glycerine at the edge of the cover-glass, when it replaces the water as the latter evaporates. Fresh specimens may also be preserved in this manner ; and, if the glycerine be subsequently replaced by water or the solution of sodium chloride, their original condition is restored.

The dark contour, the brightness of the elastic fibres and lamellæ, is but slightly diminished in glycerine ; as the refractive power of the former is considerably the greater. But, again, the peculiar lustre, exhibited by amyloid matter, the so-called *Hyaline* of Von Recklinghausen, and other colloid substances, when examined in aqueous fluids, is lost to a great extent in glycerine ; since their refractive powers differ but slightly.   In most cases, however, a careful examination suffices to disclose a somewhat clear distinction ; yet, where these substances are concerncd, it is well to first examine the preparations in water.

It has already been remarked, that small globules of fat become completely invisible in glycerine ; whence it naturally follows, that this agent is not to be employed in the investigation of tissues, which have undergone fatty degeneration.

8.—Acetate of Potassium.

A saturated solution of *potassium acetate*, as recommended by Max Schultze, may also be employed as a preservative medium, it, too, having the merit of permanency in air, not being subject to evaporation.   It has only a slight clearing effect, however, and is, therefore, best adapted for the preservation of fresh, not of hardened structures.   Particularly when the preservation of tissues which have undergone fatty degeneration is to be accomplished, this method is quite useful ; though, to be sure, the contours of the fat-globules lose their original sharpness in the course of time.

9.—Oil of Cloves, Canada Balsam.

If it be desirable to increase the transparency of the

sections, particularly after they have been deeply stained, we employ oil of turpentine, or, what is much more worthy of commendation, oil of cloves. Other essential oils have a like effect ; as, those of cedar, origanum, cinnamon, bergamot, anise, etc., besides xylol, phenol, or carbolic acid (creasote). Each one may select for himself the agent, whose odor he finds most agreeable. Inasmuch, however, as all these fluids are either not at all miscible with water, or only partially so, sections, which are to be rendered transparent, are first deprived of their water by immersing them for a few minutes in alcohol, when they readily imbibe the oil of cloves, etc. In this way the section attains a maximum of transparency, the refractive power of the above-mentioned fluids being very considerable,—much greater than that of glycerine, and nearly as great as that of glass. Even the strongest outlines of the human, or animal, elementary constituents disappear almost entirely after this treatment ; and, of unstained preparations, almost nothing remains visible.

The elastic fibres, as well, are recognized only with difficulty, and, especially, when the full illumination of the condenser (Abbé-Koch) is employed.

The colored portions of a section, on the other hand, are rendered all the more distinct. When conducting an examination by this method, therefore, we must always bear in mind the fact, that we have intentionally with-

drawn from scrutiny the majority of the structures.
Sections, prepared in this manner, are ready for mounting
and permanent preservation in the resinous mounting-
media ; and, for this purpose, Canada balsam, dissolved
in equal parts of chloroform or xylol, is principally
employed ; though a solution of mastic in chloroform,
Damar resin in turpentine, etc., are also in use.   The oil
of cloves is removed with the aid of blotting-paper ; and
its place is gradually occupied by the Canada balsam,
which has been applied at the border of the cover-glass.

STAINING: ITS PRINCIPLES, AND THE AGENTS EMPLOYED.

From year to year the art of staining has increased in
importance, and has become more and more indispensable for our purposes.   Of special import was the part
taken by it in the discovery of vegetable parasites ; and
to Weigert, P. Ehrlich, and Koch, pertain the principal
merits in this field.   The thankworthy studies of Ehrlich,
concern themselves particularly with the theory, govern-
ing the effects of dyes.

The importance of staining hangs upon the fact, that cer-
tain constituents of tissues, and even of cells, attract the
coloring matter of the solution employed with greater energy,
or in larger quantity, than the others ; and, combining with
it, these structures receive a tinge of greater or less perma-

nence. The relationship of the various substances of the human body to the different pigments, is, naturally, a most complex subject ; and yet, we have gradually discovered particular dyeing agents, or special modes of staining, for almost every tissue-component, whereby these assume an intense, specific stain, which distinguishes them from all others. *Accordingly, in many cases, staining has the value of a chemical reaction, by which we are enabled, easily and conveniently, to disclose a certain substance which lies hidden in the midst of many others ;* and, as will be readily seen, this "elective" action of coloring agents is of prime importance in the study of pathology. Elements, whose delicate outlines, masked by the investing structural maze, escape all but the closest scrutiny, become apparent, even upon a cursory examination, and, oftentimes, with a low power, immediately after this specific staining ; so that the time, occupied by the coloring process, is amply compensated by the increased convenience and certainty, thus acquired in the investigation of such objects.

In many cases staining alone affords us a clew to existing differences in the constituents of tissues and cells, which before seemed quite uniform. That the results of staining were not fully utilized before the employment of the open condenser, or the illuminating apparatus of Abbé, has been already explained.

In most instances the staining procedure is as follows :

The section is removed from distilled water into a small, shallow dish—a watch-glass, for example,—filled with the coloring solution, and so disposed that the fluid bathes it on all sides.   After remaining in this fluid for a length of time varying from a few minutes to twenty-four hours, the section is again placed in distilled water, in order that the superfluous staining matter may be washed off ; when it is examined,—either directly in glycerine, or, after its water has been removed by immersion in alcohol, in oil of cloves.   Here, the effect of the staining process is simply this : certain elements take up the color, while others remain uncolored (*election*).

Frequently, however, after being removed from the color and washed, the section is treated still further,—it is again decolorized, or partially so. That is to say, at first a completely diffused, uniform, and hence useless, staining has resulted ; but, while certain elements give up their coloring matter completely during the subsequent extractive process, other elements, which are possessed of a stronger affinity for the dye, hold it fast (Principle of maximum decolorization : Ehrlich).

This procedure, which was first employed by the author, now plays an important rôle, especially when the aniline dyes are employed.   Alcohol usually serves as the extracting agent ; in some instances, acids.

In very few cases is it advisable to attempt the staining of a section while *under the cover-glass ;* as the result is generally lacking in uniformity, is confined to the borders of the section, etc. But isolated elements, cells, etc., which are suspended in fluids, can sometimes be stained in this manner ; though, even in this case, the staining of the dried specimen, as conducted by Koch and Ehrlich, is usually far preferable. (See " Exam. of Fluids.")

In embryological and zoölogical investigations, the plan of staining organs or animals *in toto* has been followed for some time ; and alcoholic coloring solutions have been compounded, which have in view the simultaneous hardening and staining of the tissues. This offers the advantage, that the section may be transferred almost directly from the microtome to the slide, for examination ; and, aside from the brevity and convenience of the procedure, the danger of injuring or destroying the section, by the various manipulations necessary in staining, washing, etc., is greatly lessened. Still, I do not regard this method as adapted to our purposes. In the majority of cases, the task of rendering a definite opinion concerning a single specimen devolves upon us ; and we must, in consequence, retain the ability to examine the said object in as many directions as possible. Were we to adopt without question the methods of normal anatomy, we should be laboring, as it were, to the same ends ; though the conditions differ completely. The normal anatomist examines, for example, any number of eyes before deciding a scientific question ; and, to him, every normal *bulbus oculi* has an equal value,—for every variation of the mode of examination he may employ a fresh specimen. We, on the other hand, often have but a single specimen from which to ascertain the structural changes, which have taken place. To this end, we must take care to alter the preparation (by hardening, freezing, etc.) as little as possible, and to obtain a series of sections from those parts which have undergone change. Such sections may be practically regarded as equivalents. In such a case we are enabled to treat the sections in various ways, to experiment with them, in order that their alterations may with greater certainty be recognized and understood.

One can never know what surprises await him below the surface of organs, which have undergone pathological change ; whence, the specimens should be subjected to the simplest, and least disturbing, preparatory treatment before their dissection.

It is impossible, before the examination has begun, to decide definitely upon the method to be pursued ; but we must hold ourselves in readiness to follow any further lines of investigation, which are suggested by any unforeseen discoveries.

We shall not, therefore, discuss the technology of staining organs *in toto* here; but simply refer the reader, who desires to investigate this subject, to Grenacher, *Arch. f. micros. Anat.*, Bd. 16; and to P. Meyer, *Mittheilungen aus der zoöl. Station zu Neapel*, Bd. 2, 1880.

10.—IODINE.

This substance, the first staining agent to be employed in microscopical investigations, is still very often used at the present day, and, usually, in the form of Lugol's solution. Iodine is but very slightly soluble in pure water; but, upon the addition of the iodide of potassium, it becomes freely so.

The following formula may be employed :

Iodi puri	1.0
Potassii iodidi	2.0
Aquæ destillatæ	50.0

This solution may be diluted at pleasure, as occasion arises. It is to be remarked, that the stains of iodine can be preserved only with difficulty, and for a short time, in water or glycerine. The union of iodine with organic substances is not an intimate one, and it gradually evaporates; hence the color dies out. Even when the cover-glass has been carefully cemented to the slide, the coloring matter disappears after the lapse of a few years at the most.

In Canada balsam, such preparations cannot be mounted at all ; as the alcohol immediately extracts the iodine. On the other hand, preparations which have been stained with iodine seem to be well preserved in a thick mucilage.

The albuminous substances are stained a light yellow by the iodine solution, likewise the gelatinous and colloid. The cells are usually stained more deeply than the intercellular substance, and the nuclei somewhat more than the protoplasm ; accordingly, the cellular elements of fresh sections may be rendered prominent both quickly and conveniently,—for example, the cell-nests in carcinoma, etc.

The red blood-corpuscles are stained a dark brown by iodine.

A peculiar reaction attends the use of the iodine solution in the case of the following substances, particularly : Glycogen, Corpora amylacea, and Amyloid.

GLYCOGEN.

In many cartilage cells, as, for example, those of the chorda dorsalis, the proliferating layer of epiphyseal cartilage, even in the normal state, but particularly in the cells of rachitic cartilage, and those of enchondromata, an intense wine-red stain, affecting the whole or a part of the cell, is produced by the iodine solution.

The reaction is best shown, when the remaining

structures are only faintly colored. The red stain is due to the presence of *glycogen* in the cells, as shown by Neumann and Jaffé ; while, without the iodine staining, parts which are rich in glycogen exhibit a homogeneous brilliancy.

The cells of the chorionic tufts, and of other embryonic structures, react in the same way with iodine ; as shown long ago by Claude Bernard. Bock and Hofmann likewise made use of iodine staining, when studying the variations in the amount of glycogen contained in the liver cells. In normal stratified pavement-epithelium, also, as well as in carcinomata of luxurious growth, Schiele—a pupil of Langhans—found glycogen in abundance. The fact, that a large part of the glycogen is extracted from the cells of sections which are placed in watery solutions (in alcohol, as we know, glycogen is insoluble), was recently discovered by Ehrlich ; and, in consequence, this substance naturally escapes observation. To prevent this solution or diffusion of the glycogen, he adds mucilage of gum arabic to the iodine solution, making a thin fluid, in which the preparations are both examined and preserved. Employing this method, Ehrlich found that in *diabetes* glycogen, in large quantities, regularly appeared in the epithelial cells, lining the renal uriniferous tubules, and, particularly, in the boundary layer between the cortical and medullary structures.

CORPORA AMYLACEA.

Starch (amylum) granules, as is well known, are stained a deep blue color by iodine, a fact which renders their detection certain, when they are present in the contents of the stomach, intestine, or buccal cavity, or when they occur as a chance impurity.

The so-called corpora amylacea, of almost regular occurrence in degenerative processes of the nervous system and in the white substance of the brain and cord of elderly individuals, possess a certain, but purely external resemblance to starch granules ; though they are to be regarded as degenerative products of the medullary sheaths. Iodine imparts to these a deep wine-red color.

Furthermore, certain concretions, which are sometimes found in the lungs and very frequently in the prostate gland, are also designated as corpora amylacea. These colorless or yellowish-brown bodies are characterized by a concentric striation, and by a more or less intense stain, which is imparted by iodine,—the color varying from wine-red to brownish-black. As to their nature and significance, little is known ; but they are entirely distinct from starch or amyloid.

AMYLOID.

The *amyloid substance* is characterized by the wine-red color, which it assumes when exposed to the action of iodine. And here, too, quite weak solutions, of "cognac-color" (*Cognacfarbe*), are best employed ; for, though the reaction ensues more gradually, requiring several min-utes, the result is clearer and more beautiful.

In many cases of amyloid degeneration the color changes, ranging from a dark-green to a blue, when sul-phuric acid is added. Having been slightly stained—to a light yellow—with iodine, the section is placed in a one-per-cent. solution of sulphuric acid ; when the reaction occurs, either immediately, or in the course of a few minutes. As already stated, this blue or green stain, by which the altered structures are contrasted with their surroundings much more vividly than when iodine alone is employed, does not appear in all instances of amyloid degeneration,—the sulphuric acid frequently producing no change of color, but simply intensifying the brown-red tone.

Furthermore, one finds in many cases of amyloidal change, that some of the degenerated elements become green, or blue, with iodine and sulphuric acid, while others asume only a dark-red color. For example, in several cases, now before me, the arteries and vasa afferentia of the kidney became dark-red ; while a deep blue distinguished the Malpighian bodies. This picture, which is extremely striking by reason of its regularity, is but seldom seen, apparently ; though, in the case of amyloid arteries for example, one frequently observes that the greater part is tinged a red color, while only a few scattered patches of blue are visible.   It is possible, that these variations in color depend upon the *age of the amyloid*, the more recent places being stained red by iodine and sulphuric acid ; the less recent, blue.   Thus, I have often seen the arteries and capillaries of the greatly enlarged splenic corpuscles turn blue ; while the vessels of the less diseased, and hence presumptively younger, pulp were stained red.   Further distinctions between these two kinds of amyloid have not, thus far, been discovered.

Both without staining, and when the aniline dyes are employed, the amyloid exhibits only a homogeneous quality ; where treatment with iodine and sulphuric acid discloses the well-marked distinction in color.

The characteristic reaction with iodine has thus far proved to be a necessity, in the microscopical diagnosis of amyloid.   Other colloid and hyaline substances possess its homogeneous, glistening appearance, but are stained only a light yellow by iodine.   The red color, which results from treatment with violet aniline dyes, is not always characteristic of amyloid, as it appears.   Casts of the tubuli uriniferi, for example, are often stained red by aniline violet ; while iodine imparts to them merely a yellow tone, which, in case the action be too strong, merges into brown.* From this, it is to be concluded, that casts which are stained red by aniline are not made up of amyloid proper : perhaps they represent an early stage of this substance.

The chemical nature of amyloid has been the subject of much study ; and it is known, that it is a substance, rich in nitrogen, and allied to the albuminous bodies. On the other hand, the cause of the reaction with iodine, either alone or combined with sulphuric acid,-- a reaction from which it has received its very inappropriate name,—is completely unknown. We are quite ignorant concerning the chemical nature of the red iodine combination, or of the blue structures into which sulphuric acid transforms the same.

It is to be observed, that cholesterine, also, takes quite a dark color in a weak iodine solution ; and, if a drop of strong sulphuric acid be permitted to pass under the cover-glass, a beautiful blue color is likewise imparted to the edges of the tablets.

II.—CARMINE.

The introduction of carmine-staining dates from the year 1858, and was due to the labors of Harting and Gerlach.

---

* In certain cases, however, the casts assume a reddish-brown shade with iodine.

a. Ammonia-Carmine.

In the original formula of Gerlach, the carmine was dissolved in ammonia, about one part of finely pulverized carmine being dissolved in one part of the strong liquor ammoniæ and from fifty to one hundred parts of water. The mixture is exposed to the air for twenty-four hours to permit the evaporation of the greater part of the ammoniæ, after which it is filtered. The less free ammonia contained in the solution, the more admirable its effects ; but its liability to mould, renders its frequent renewal necessary.

The carminate of ammonia stains a large number of the substances occurring in the animal economy, and very quickly, too ; while the stain is real, or permanent, particularly, when, after a thorough washing, the section is placed in a weak solution of acetic acid, for the latter agent's fixing or mordant effect.

If the washing be not effectual, the section will be completely ruined by the granular carmine precipitate, which forms in the acetic-acid solution.

This agent tinges the protoplasm and nuclei of almost all cells, the fibrillar basement-substance of connective tissue, smooth and striped muscular fibres, the matrix of osteoid tissue and of decalcified bone, fibrine, the neuroglia, the axis-cylinders of nerves, most colloid substances, etc.; while the matrix of hyaline cartilage, elastic tissue, the corneous structures, the medullary sheaths of nerves, fat, mucus, calcined bone, etc., remain uncolored.

Ammonia-carmine is principally used in the study of the nervous system, to demonstrate the axis-cylinders.

Nervous tissue is hardened, for the most part, in solutions of the chrome salts, as we have already said ; and the time required for the staining is proportional to the length of time the specimen has remained in the hardening fluid, several days often passing before a color of sufficient intensity is obtained. And, particularly, when very small axis-cylinders are to be investigated, as those of the optic nerve, which require a very deep stain, this drawback makes itself felt. In these cases the tingeing may be intensified, or hastened, by placing the carmine solution in an enclosure, heated up to 50° C., or 122° F. (Obersteiner); when an hour, or even less, suffices to secure the desired result.

Henle and Merckel have given us another method, deserving of great commendation : the section is first placed in a solution of the chloride of palladium (1–500) for about ten minutes ; when it becomes straw-colored ; after which it comes into the carmine solution. Only a few minutes now elapse before it is stained a deep red ; and, after being carefully washed in water, dehydrated in alcohol, and rendered transparent by the oil of cloves, it may be examined. By this method the medullary sheaths are stained yellow ; the neuroglia, ganglion cells, and axis-cylinders an intense red.

If it be desirable to render the nuclei conspicuous also, this may be affected by placing the section, already stained by carmine, in a solution of hæmatoxyline. (See page 76.)

These double stainings are to be particularly recommended for the investigation of secondary degeneration of the spinal cord, sclerosis, atrophy, etc. The diffuse red stain of the neuroglia, however, is frequently a disturbing feature in preparations thus treated ; and, to remove this, Ranvier lays the sections, after the carmine staining, in an alcoholic solution of formic acid (formic acid, one part ; alcohol, two parts) for five to ten hours ; when axis-cylinders and nuclei remain red, while the neuroglia is decolorized. (*Compt. rend.*, 1883, Nov.)

Ammonia-carmine is, moreover, largely employed in the examination of the *osseous system.* Here, too, double staining (ammonia-carmine and hæmatoxyline) can be employed with advantage.

In the investigation of rachitis and osteomalacia in particular, is this accentuation of the osteoid tissue, by the use of ammonia-carmine, of real value ; and, for this purpose, it is best to make use of fresh preparations, *i. e.* such as have not undergone artificial decalcification.

Many modifications of carmine-staining have been

proposed and recommended ; but we shall here mention only a few, which have a special value in the study of pathological anatomy.

b. Picro-Carmine.

(*Schwarz, Ranvier.*)

The picro-carmine of the shops, is generally useless ; and Weigert advises the addition of a small quantity of acetic acid to this, in order to obtain a good staining agent. Should a precipitate occur, this can easily be dissolved with a trace of ammonia.

The author prepares a picro-carmine, which stains very rapidly, after the following recipe : From two to four parts of a saturated solution of picric acid are slowly added to one part of ammonia-carmine (1 carmine, 1 ammonia, 50 water), the mixture being continuously stirred. The picric-acid solution is finally added drop by drop, till the precipitate, which forms at first, is no longer dissolved by stirring. The amount of picric acid necessary is proportional to the quantity of ammonia present. The mixture is now filtered, and for each 100 c. c., a few drops of carbolic acid are added, to prevent decomposition of the fluid. Any cloudiness which may subsequently ensue is dispelled by a few drops of liquor ammoniæ.

This staining fluid is very serviceable in the majority of

instances, as it effects a *double staining* in a few minutes, the nuclei becoming a *deep red*, the fibrillar substance of the connective tissue, etc., assuming a *light-red shade ;* while the protoplasmic structures, the smooth and striped muscular fibres, the cornified elements, most hyaline and colloid substances, etc., are stained a *more or less deep yellow.*

The contrast often becomes still more striking, if, after the staining process, the sections are placed for half an hour in glycerine, to which hydrochloric acid has been added (one part of hydrochloric acid to one hundred parts of glycerine). The picro-carmine solution stains more rapidly and deeply when it contains some free ammonia ; but, in such a case, the carmine stain at first predominates, and only by treatment with the acidulated glycerine is the red dye removed from the protoplasmic and intercellular substances, etc., thereby exposing their yellow picric tinge (Neumann). At the same time, however, the red dye is fixed in the nuclei. It remains to be said, that the red nuclear stain is permanent in neutral or acid fluids ; while the yellow picric stain disappears quite early. To preserve the yellow shade in the preparations, it is customary to add picric acid to the water, glycerine, and alcohol, which are to be employed ; but only in quantities sufficient to produce in them a light-yellow tinge. When this is done, the picro-carmine stains may be well preserved, both in glycerine and in Canada balsam. On account of the ease with which the double tinging is produced,—the sharp prominence of the nuclei, on the one hand ; on the other, of the protoplasm, hyaline substances, corneous tissue, smooth muscular fibres, etc.,—a special value attaches itself to picro-carmine. The discovery of tubercles in scrofulous granulations, in tissues affected with lupus, etc., has been materially facilitated, in the author's case, by the employment of this method ; while, for the nervous and bony structures, many glands, etc., picro-carmine may be highly commended, also.

c. **Borax-Carmine**
(*Grenacher.*)

Carmine	0.50
Borax . . .	2.00
Distilled water	100.00

These are mixed in a porcelain dish, and heated to the boiling point. While the bluish-red fluid, thus produced, is being stirred continuously, diluted acetic acid (about five-per-cent.) is added, drop by drop, until the color of the mixture changes into that of the ammonia-carmine solution. After standing for twenty-four hours, it is decanted and filtered. As a preservative, a few drops of carbolic acid are added.

A section, which is placed in this solution, is deeply

stained, in the course of a few minutes ; but, in consequence of the diffusion of the color, it becomes practically useless. On the other hand, *the most beautiful, isolated, nuclear staining* is obtained, if the section be now placed in a receptacle, filled with hydrochloric acid, alcohol, and water, in the following proportions :

Hydrochloric acid   .   .   1.00
Alcohol 70.00
Distilled water 30.00

In this fluid, the section immediately surrenders a portion of its coloring matter, which invests it in the form of a red cloud ; and, after the lapse of from a few minutes to half an hour, the specimen is washed in water or alcohol, and examined in glycerine or oil of cloves.

This method affords the *most intense nuclear staining ;* but, in employing it, the action of the hydrochloric acid must be taken into consideration, and allowed for ; as, solution of the lime salts, swelling of the fibrine, protoplasm, fibrillar substance, etc.

d. Alum-Carmine.

(*Grenacher.*)

One gramme of carmine is warmed up with 100 c. c. of a five-per-cent. solution of alum ; and the mixture is permitted to boil for twenty minutes. After cooling, it is filtered.

By the use of this fluid one obtains, in from five to ten minutes, an almost purely nuclear tingeing ; though this is not quite so intense, as that afforded by the preceding borax-carmine.

e. Alum-Cochineal Fluid.

(*Partsch and Czokor.*)

One part each, of the finest cochineal (of which carmine is a derivative) and alum, are boiled down, with 100 parts of water, to half the original volume ; when a trace of carbolic acid is added, and the fluid is filtered.

The effect of this staining fluid is quite similar to that of the preceding alum-carmine. I esteem it, when a simultaneous staining of the nuclei and axis-cylinders, in sections of the cerebro-spinal axis, is desirable, after previous hardening in solutions of the chrome salts. The staining takes place within twenty-four hours ; when the nuclei are found to have a color shading more into the violet, than that of the axis-cylinders.

f. Lithium-Carmine.

(*Orth.*)

Two and one half parts of carmine are dissolved in one hundred parts of a saturated solution of the carbonate of lithium.

In this fluid, sections are stained in a few minutes ;

and, atter partial decolorization in acidulated alcohol (see sect. *c*), magnificent nuclear staining is the result.

Picro-lithium-carmine is prepared by adding two to three parts of a saturated solution of picric acid to the preceding staining fluid.

This agent, also, stains very quickly, and affords the advantage of double tingeing, just as does picro-carmine. Here, too, a subsequent partial decolorization can be effected in glycerine, or alcohol acidulated with hydrochloric acid ; whereby the colors are more vividly contrasted. Inasmuch as this fluid keeps very well also, we feel justified in indorsing it highly.

12.—HÆMATOXYLINE.

Weigert's Method of Staining the Cerebro-Spinal Centre.

Staining with hæmatoxyline, is one of the most certain and excellent modes of rendering the nuclei of cells markedly conspicuous. The hæmatoxyline crystals dissolve readily in alcohol, forming a brownish tincture. If a small quantity of this be added to a watery solution of alum, a blue-colored fluid is produced in a few minutes, which only attains its full staining power in the course of a few days. At the same time, or soon afterward, a granular separation of the dye also begins, which is detrimental to preparations treated with this solution ; accordingly, it must always be freshly filtered before use.

To obtain a lasting fluid, of constant staining capacity, the following formula may be employed :

Hæmatoxyline .	.	.	.	.		2.00
Alcohol .	.	.	.	.	.	100.00
Distilled water .	.	.				100.00
Glycerine .	.	.			.	100.00
Alum	.	.	.	.	.	2.00

If occasion arise, a little acetic acid may be added to avoid over-staining (P. Ehrlich). It is to be remarked, that this fluid only reaches its full staining power eight days after its preparation.

If a section be laid in this brown fluid, it is very quickly stained brown, also ; but, after being washed in distilled water, this color changes in a few minutes to blue. We then find, that the stain is *almost entirely confined to the cell nuclei and (most) bacteria*, which may be present. Should other tissue elements be stained at the same time, acidulated alcohol is to be employed for its decolorizing effect (see art. 11, sect. *c.*), the nuclear stain remaining well preserved. For many purposes, it becomes desirable that the protoplasmic substance be likewise stained ; and this may be effected by the subsequent use of a saturated solution of picric acid, or with eosine (see *Eosine*).

From the fact that preparations stained with hæmatoxyline gradually lose their color in glycerine, they are best preserved in Canada balsam.

Quite recently, Weigert has acquainted us with a very valuable method of employing hæmatoxyline staining for the cerebro-spinal centre, by which we are enabled to exhibit, in an elegant manner, those fine medullated nerve fibres which could formerly be demonstrated only with the greatest difficulty.*

This quite peculiar process is somewhat complicated. The specimens are hardened in Müller's or Erlitzki's fluid, and, without a previous soaking in water, are immediately placed in alcohol. The sections, also, must not be kept in water before staining, but in alcohol. The staining fluid consists of hæmatoxyline, one part ; alcohol, ten parts ; water, ninety parts. This mixture is boiled, and allowed to stand for several days. The sections are exposed to the action of the staining fluid for from one to two hours, and, preferably, at a temperature of about 40° C. (104° F.) ; hence a hatching-oven is employed. After being rinsed in water, the deep-black, strongly overstained sections are, for the most part, decolorized in a mixture containing two and a half parts of the ferrocyanide of potassium, and two parts of borax, in one hundred parts of water. The solution is permitted to act from half an hour to an hour, or until the gray substance appears of a yellowish tinge,—the white substance remaining black. The sections are now well

---

* *Fortschr. d. Med.*, Bd. II. S. 190.

rinsed in water (the potassium ferrocyanide is precipitated by alcohol), to be afterward placed in alcohol,
xylol, and Canada balsam successively.

The white substance (medulla) of Schwann is sharply
distinguished by the dark tinge, which is imparted to it
by this process ; while the axis-cylinders, ganglia, cells,
nuclei, etc., remain almost colorless. The long postu-
lated alteration of the gray matter of the cord in *tabes*,
or *locomotor ataxia*, which before had been sought for in
vain, has confidently been demonstrated by the aid of
Weigert's method. The alteration consists in the atrophy
of the fine, reticulated nerve fibres within the columns of
Clark. This result is of great theoretical importance ;
and, inasmuch as similar changes perhaps remain to be
discovered (in the cortex cerebri in progressive paralysis,
in secondary degenerations, in retinal affections, etc.), we
have described the process at length. Sections prepared
by this method are extremely profitable. Supplementary
nuclear tingeing may easily be effected with alum-
carmine, for example.

13.—EOSINE.

Eosine forms a fluorescent solution, which appears
rose-red by transmitted light ; while reflected light gives
it a greenish hue. Even in the strength of 1–1,000, it imparts a deep rose-red stain to sections in a few minutes.

The coloring is, for the most part, very diffuse, affecting the most various structures. Even the red blood-globules, in sections taken from alcohol preparations, are stained an intense rose-red, which is still more intense, when chrome salts have been used in hardening. Absolute alcohol removes the color very quickly at first, more slowly later ; so that, with some attention, this agent can easily be employed to produce any desired shade of red, Staining with eosine alone is only very seldom advantageous (concerning the eosinophilous cells of the blood, see below) ; while it is very often employed in double staining, with agents possessing an affinity for nuclei. For the sake of contrast, those agents are best employed for this purpose, which stain the nuclei blue ; as gentian- or methyl-violet, or hæmatoxyline. Following the example of Renaut, a mixture of eosine and hæmatoxyline may be prepared, by which the double staining is accomplished simultaneously ; and it is only necessary to add about 0.5 of eosine to the formula, given for the hæmatoxyline solution, in order to produce such a double-acting fluid. At first the sections generally exhibit too strong an eosine stain ; but, after a short bath in alcohol, the proper shade appears, when the section is generally best examined in oil of cloves.

In very many instances this method of staining sections is the best and most convenient. The nuclei of the lymphoid cells generally appear the deepest stained, then those of the capillaries, other endothelial cells, the connective tissue ; less deeply, the epithelial nuclei, etc. Variations in the intensity of stain also appear in the protoplasm, rendering the recognition of the individual elements possible, even with low powers. For example, the investing cells of the gastric follicles, the giant-cells of tubercle, etc., take an especially deep color.

14.—ANILINE BLACK (NIGROSINE)—ANILINE BLUE.

These two agents have a very similar action, and are used for staining the axis-cylinders, in sections of nervous tissue. The section is placed in a solution of about one per cent. for a few minutes, when it takes a very dark tinge. It is subsequently washed in alcohol, and thus deprived of most of its color. When the proper shade is obtained, which can easily be ascertained with the microscope, the section is placed in oil of cloves or Canada balsam. The axis-cylinders, ganglion cells, and, in a less marked degree, the neuroglia disclose a very serviceable blue or black tinge.

They can also be employed to stain protoplasm. For example, even quite weak solutions impart a very dark and characteristic stain to the investing cells (the former so-called peptic cells) of the gastric follicles.

15.—The Nuclear-Staining, Basic Aniline Dyes.

The following basic aniline dyes, employed because of their special affinity for nuclear structures, are most used : * Vesuvine (Bismarck-brown), fuchsine, gentian-violet and methyl-violet, methyl-blue ; besides methyl-green, dahlia, magdala, etc.

These various substances deport themselves very similarly towards the tissues, and may, therefore, be considered together. They are all quite soluble in water and alcohol ; but watery solutions are employed by us almost exclusively,—alcoholic only seldom. It is best to keep constantly in stock a concentrated aqueous solution of one or more of these agents, in which there is an excess of coloring matter. Immediately before using, the amount needed is passed through a filter. It is often convenient to have, instead of this opaque, deeply-colored solution, a weaker and less concentrated one— about 1 to 100—on hand, to which about ten per cent. of alcohol may be added for its preservative effect.

---

* These staining agents may be obtained either directly from dealers in dye-stuffs, or through those opticians mentioned in Chapter I., Art. 3.—S. Y. II.

Nuclear Staining.

(*Non-nucleated cells.*)

If a section be placed in one of these solutions for a short time—a few minutes,—it becomes very deeply tinged ; and, if it be washed in water and examined, an almost universally diffused, uniform, and hence useless, coloration is found. Only after the action of alcohol, do the advantages of the staining become apparent. When placed in a small dish containing alcohol, the section gives up a large amount of coloring matter, which forms a floating cloud about it ; and when, after the lapse of a few minutes, it is again placed under the microscope, the beautifully distinct color of the nuclei, etc., is remarked. As a general thing, the nuclei of the lymphoid cells, as well as those of the connective tissue and endothelial cells, are much more deeply stained than the epi-thelial nuclei. Almost all of the remaining substances are still uncolored ; but there are some few exceptions, which will be considered presently.

The process, accordingly, is a short one : the section remains in the dye-liquid for a few minutes, then in alcohol for a like time ; whence it is transferred either to oil of cloves, which renders the unstained structures nearly transparent and ready for mounting in Canada balsam ; or, on the other hand, it is again placed in distilled water, and examined in glycerine.

It is to be remarked here, that the colors are well preserved, when the sections are mounted in balsam ; while,
in glycerine, the nuclei retain their color only when Bismarck-brown, or vesuvine, has been used,—the stain of
methyl-violet, etc., being gradually obliterated in this
medium (Weigert).

For this reason, and for the majority of purposes,
staining with Bismarck-brown is to be specially commended, and particularly, too, when the sections are
fresh ; as these produce no precipitation in solutions of
Bismarck-brown, while, in a solution of methyl-violet,
they effect a troublesome granular deposit.

By means of this distinct, isolated nuclear staining, which
may be produced not only by the aniline dyes mentioned
but also by hæmatoxyline, picro-carmine, borax-carmine,
lithium-carmine, etc., Weigert was led to the important
discovery, that, in a large number of pathological processes, the cell nuclei disappear, either universally, or at
least in certain categories of cells,—a fact which thereto-
fore had remained practically unknown. The absence
of nuclei was first detected in the cells of the deep layers
of the epidermis in small-pox ; then, in diphtheria ;
again, in various organs in the neighorhood of colonies
of micrococci ; in the convoluted uriniferous tubules, in
cases of chrome poisoning ; in renal infarcts ; cheesy degenerations, etc.

It soon appeared, that the absence of nuclei occurs in necrosed cells, that remain exposed for a certain time after death to the action of the nutritive juices, which circulate, though in a lessened degree, through the lymph lacunæ. In many instances, the protoplasm of the cells assumes at the same time a brightly glistening, homogeneous appearance ; and these are the cases for which Cohnheim introduced the expression "coagula- tion-necrosis." This term has since been often misap- plied, and should, in my opinion, be limited to those cases in which coagulation, in addition to necrosis, has been demonstrated, or at least rendered probable. In any event, every cell, upon the nucleus of which a specific staining procedure does not act properly, should not be regarded as a victim of "coagulation-necrosis." It should be stated of aniline stainings, that the nuclei of quite normal epithelial cells may sometimes appear perfectly colorless, when the staining ensues in a solution of acid reaction, or when dilute acetic acid is employed as a decolorizer in addition to alcohol. Moreover, it has by no means been proved, that every cell, the nucleus of which cannot be demonstrated, is to be regarded as necrotic ; and, hence, in such doubtful cases of cells without colorable nuclei, it is as well not to speak im- mediately of "coagulation-necrosis."

Before the complete disappearance of the nucleus,

small, deeply stained granules are often found in its place, which may indeed be regarded as the remains of the disintegrated nucleus, though often mistaken for micrococci by beginners. Their variable size, however, amply distinguishes them from these microbes.

Besides the nucleus, this method stains :

1. The matrix of hyaline cartilage.

2. Mucous substances, and particularly the mucous contents of glands.

3. Most forms of the micrococcus and bacillus, microbes which we shall soon consider further.

4. Certain protoplasmic granulations,—the protoplasm of the so-called nutrition cells.

NUTRITION CELLS.

(*Mastzellen.*)

These so-called nutrition cells, the objects of Ehrlich's careful study, are nearly spherical, though sometimes flat and spindle-shaped, structures about twice the size of lymphoid cells. They consist of a quite coarsely granular protoplasm, whose granules are deeply stained by the basic aniline dyes employed in the manner described above. When the violet dyes are used, the granules assume a reddish shade ; while the nucleus remains unstained, and appears like a bright spot in the midst of the strongly stained particles of protoplasm. The cells are distributed throughout the *connective tissue*, but are especially numerous in mucous membranes and in sub-mucous tissue. The intermuscular tissue, serous membranes, etc., also contain them, though nearly always isolated from their fellows, and lying for the most part in the vicinity of vessels. Very little is known as yet of their physiological or pathological significance ; while their connection with nutrition cannot be demonstrated. They are present in large numbers in slowly growing formations of connective or granulation tissue ; as in elephantiasis, and in the neighborhood of tumors.

They have also been discovered in leucæmic blood, but are wanting in normal human blood.

Amyloid Staining with Violet Aniline Dyes.

The basic aniline dyes, methyl-violet and gentian-violet (Leonhardi's fluid which was formerly sometimes employed probably contains one of these substances), exhibit a peculiar reaction with amyloid substances,—as Heschl, Juergens, and Cornil discovered almost simultaneously,—staining them a deep red ; while the nuclei, etc., are colored blue. The red color of the amyloid is unaffected by dilute acids, but in alcohol it is immediately extracted (the light reddish tinge of the nutrition cells is preserved in alcohol).

Accordingly, preparations which are stained in this way cannot be decolorized with alcohol in the ordinary manner ; but a dilute acid, as a one-per-cent. solution of acetic acid, is employed for this purpose.

In the first place, however, the objects are not tinged too deeply, a more dilute staining fluid—about 1 to 1,000 —being employed ; but, even in this strength, a few minutes suffice for the operation. Glycerine is employed for the examination and mounting of the sections ; though it is true, that in this medium the nuclear stain disappears in time.

The distinction between the red amyloid and the remaining blue portions of the object is very apparent ; besides which a distinct blue nuclear tingeing is present in well-prepared specimens. As a consequence, this method possesses great advantages over the only one formerly known for staining amyloid ; viz. iodine, or iodine and sulphuric acid.

By a systematic employment of this method, Eberth, as we know, arrived at the important conclusion, that the cellular elements of the liver, kidneys, and spleen, for example, as well as the smooth muscular fibres of vessels, are never the subjects of amyloid degeneration, but that this always affects the intercellular substance and homogeneous membranes only. For the decision of these questions the aniline method is decidedly superior ; but caution must be exercised, for the reason that every red tinge caused by methyl-violet does not of necessity indicate genuine amyloid degeneration. Other hyaline formations also, perhaps closely allied to the amyloid substance, such as certain urinary casts, for example, exhibit the reaction, though they cannot be regarded as amyloid. The two varieties of amyloid, which, as we have seen, are disclosed by the iodine and sulphuric-acid reaction, receive a uniform tinge when the aniline dyes are employed.

Curschmann has recently recommended methyl-green as a staining agent for amyloid, as this renders the amyloid parts violet ; the normal substance, and particularly the nuclei, green. The difference in color is certainly very striking ; but the violet tinge of the amyloid is due simply to the chance admixture of methyl-violet in the impure methyl-green of the shops. As a general thing, the employment of impure manufactures is not to be recommended for our purposes ; but we should rather cleave to definite chemical substances in the purest state.

DEMONSTRATION AND STAINING OF BACTERIA.

THE staining of bacteria has played a prominent rôle in the important discoveries relating to infectious diseases, which have been made of late years ; and, therefore, we need no excuse for considering the subject more at length than the limited scope of this little work would seem to warrant.

a. Demonstration of Unstained Bacteria.

To demonstrate unstained bacteria in the tissues, we usually avail ourselves of a quality, by virtue of which these microbes are able to resist the action of acids and alkalies. A section, taken from a fresh specimen or from one hardened in alcohol,* is rendered almost wholly transparent by the action of strong acetic acid or a dilute—say two-per-cent.—solution of caustic potash or caustic soda ; and, amid the few elements which resist this treatment, the bacteria immediately become prominent :

1.—*From the Characteristic Form of the Individual Organisms.*

This is especially true of the bacillar forms, which, at the most, can only be confounded with diminutive crystals. Thus, the bacilli of typhoid fever and tuberculosis can be very well demonstrated in the organs by this method ; though it is best to employ specimens which

---

* Hardening in chromic acid, Müller's fluid, etc., is not adapted to the study of bacteria ; inasmuch as this acid and its salts occasion abundant dark granulations in the tissues, which are cleared up with difficulty.

have been in alcohol only a short time, as, otherwise, the sections do not usually become thoroughly transparent. Klebs and Baumgarten have laid decided stress upon this point.

However, a sufficient degree of clearness can almost always be secured, even in the case of old alcohol preparations, by mounting the specimen in liquor potassæ or acetic acid, and heating the slide till ebullition begins.

2.—*From their Characteristic Groupings.*

The individual microbes become joined together in linear groups of two or more, forming diplococci, streptococci, streptobacteria, etc., or are aggregated into colonies—gliacoccus, etc.

In such cases the possibility of confounding the microbes with inorganic precipitates will very seldom arise ; while the fact that the micro-organisms resist the action of ether and chloroform, serves to distinguish them from minute fat-globules.

The recognition of the chains and groups of micrococci is facilitated by the peculiar lustre of the individual organisms ; by their brownish color, which however appears only in the case of certain varieties, and even then only when they are present in several superimposed layers ; and by their approximately homogeneous nature, and sharply defined boundaries under high powers.

Absolute certainty as to the organic nature of these objects is afforded, when their growth can be demonstrated. This can be done, when bacteria are developed in the lumina of vessels. Here they distend the vascular parietes in an irregular manner, by reason of their variable growth and multiplication, thus forming *varicose expansions.* This frequently occurs in cases of metastatic pyæmia, endocarditis ulcerosa, etc. ; when the blood-vessels, capillaries, and small veins are found to be thus altered in form. The author found a similar varicose injection of the lymphatic vessels by bacteria in acute croupous pneumonia. These forms of "capillary embolism," of irregular distension of capillaries with granular material, had long been observed ; it was already known that these emboli became the foci of inflammation ; yet, this granular material was designated detritus, until Van Recklinghausen, and soon afterwards Klebs, Waldeyer and others, made the important discovery that the "detritus granules" represent living, parasitic organisms —micrococci, basing their views essentially upon the varicose form of the injection. For only a substance which is capable of proliferation can produce such an unequal, nodular form of vascular injection. This fact once proved, it naturally became superfluous to demonstrate the phenomena of growth in every individual instance, ere making a diagnosis of "micrococcal colonies."

But if we find heaps or chains of small granules in a section cut from a fresh specimen, or from one hardened in alcohol ; if these granules are nearly uniform in size, and resist treatment with alcohol and ether, as well as the energetic action of concentrated acetic acid and the alkalies, even when seconded by heat,—then we are justified in pronouncing the granules to be organisms.

In all important and any wise doubtful cases, one naturally must not fail to employ the staining reaction, in the manner about to be described. The discovery of bacteria in sections is also greatly facilitated by these staining procedures.

The methods employed for the *cultivation* of microbes will not be considered here ; but, for particulars, the reader is referred to the classical exposition of Koch in the "Communications of the Imperial Board of Health," 1881.

b. Staining of Micrococci, etc.

Bacteria, and the nuclear substance of cells, are affected by dyes in a very similar manner ; but the former are markedly distinguished by their resistance to alkalies, in which the nuclei are immediately dissolved. The methods followed in nuclear staining may accordingly be employed for most forms of bacteria ; and, inasmuch as we here have to do with very minute bodies, our efforts should be directed toward the production of the *most in-*

tense staining effect possible. For this purpose, hæmatoxyline and the nuclear-staining carmine dyes are less fitted
than the basic aniline dyes (Weigert). Quite concentrated aqueous solutions of these latter are accordingly
used,—a considerable addition of alcohol injuring the
staining power of the solution.

According to Koch, the tingeing may be further intensified by conducting it in a closed receptacle, the temperature of which is raised to about 50° C. (122° Fahr.).

The staining occupies from a few minutes to half an
hour ; and the subsequent decolorization in alcohol—
which must be pure, and, above all things, free from acid
reaction—should not be too long continued, inasmuch
as the color of the bacteria also is sometimes gradually
diminished. For examination, the specimen is always
mounted in oil of cloves or Canada balsam, and should
be illuminated by an open Abbé condensing apparatus.

The selection of one of the previously mentioned dye-
stuffs is generally a matter of little importance. Preparations which are colored with vesuvine, or Bismarck-
brown, can be photographed ; on the other hand, the
blue or red tingeing with gentian-violet or fuchsine usually affords the most vigorous and intense pictures for
ocular demonstration.

The stained bacilli are unmistakable ; but the micrococci might be confounded with the remains of disinte-

grated nuclei (see " Nuclear Staining "), or with the granules of nutrition cells. Their varying size, however, and the fact that they occupy the site of the original nuclei, generally serve to distinguish the former ; while the latter are also easily recognized, as they lie grouped about the centrally disposed nucleus, which has remained unstained. Yet, the recognition of these things is attended with a succession of apparently insuperable difficulties for one unaccustomed to accurate observation. I have often known an investigator to be disposed to regard as micrococci, stained mucous coagula of irregular shape, or other purely accidental substances. Others permit their sections to lie for days in impure water instead of alcohol ; when bacteria in abundance collect upon their surfaces and edges, occasioning much perplexity, if the sections be subsequently stained. There are, indeed, natures which never, or only after long and painful training, become competent to undertake histological investigations of the finer order.

Now, are all parasitic micro-organisms colored in this manner ? No, such is not the case. One might as soon expect a universal method for histological investigations in general, as a *single* mode of demonstrating these microbes. In the first place, however, the micrococcus, and, so far as is known, all its varieties, are stained in the manner described. Only this restriction must here be made, that the micrococci usually lose the power of receiving or of retaining the coloring matter after death ; and one quite frequently finds, in addition to deeply stained aggregations of micrococci, much paler, and even quite colorless, granular collections in the interior of organs, which in all probability may be regarded as lifeless micrococci. In the moniliform arrangement of micrococci also, variations in the intensity of the color of individual members of the chain are observed, which admit of a like interpretation. Nevertheless, it is possible that further study will disclose the fact, that the various species of micrococci differ in their color receptivity ; though, as yet, nothing of this sort is known.

The micrococci of malignant endocarditis, pyæmia, erysipelas, gonorrhœa, etc., deport themselves quite uniformly toward the method employed by us.

The micrococci of pneumonia, as we know, are characterized in many instances by a peculiar capsule, which can be only slightly stained with gentian-violet and fuchsine ; while the coccus itself takes a very dark tinge. Bismarck-brown and methyl-blue impart a nearly, or quite, uniform stain to both capsule and coccus.

The spirilla of relapsing fever have a special place. Most investigators have completely failed in their endeavors to demonstrate these *in situ* in the tissues. As is known, the spirilla are differently constituted from

most of the other forms of bacteria, being quickly destroyed by acids and alkalies,—yes, even by distilled
water. In this they resemble protoplasm rather than
nuclear substance, and, accordingly, cannot for the most
part be stained by the processes usually employed for
nuclei. R. Koch alone, the greatest master in this field,
by employing brown aniline dyes has succeeded in staining these organisms in the *substance* of organs also, and
in photographing them. He declares, that even the demonstration of spirilla in hardened tissues is a difficult
task.

The *bacilli* deport themselves in various ways. The
forms which usually appear during cadaveric decomposition, as well as those of anthrax, take a deep stain ;
while it was at first denied, that the bacilli of typhoid
could be tinged at all. This, however, is not true. At
the most the typhoid bacilli are tinged somewhat less
deeply ; and even this slight distinction completely disappears, when the staining is performed in a heated
enclosure. On the other hand, small unstained points
are found within these microbes,—apparent solutions of
continuity ; partly round, partly elliptical, and taking up
about half the width, and over, of the bacillus. Possibly, these are the spores of the bacilli ; possibly, "vacuoles." According to Gaffky, another kind of spore
formation occurs in the bacilli of abdominal typhus,

when cultivated at the temperature of the body—viz., terminal spores, of the width of the bacillus.

The bacilli of lepra are colored by gentian-violet, methyl-violet, and fuchsine, but not by Bismarck-brown ; and one finds in them also unstained spots, similar to those of the typhoidal bacilli.

c. Gram's Method.

While both bacteria and nuclei are stained by the method already described, and which until recently was the only one employed for these microbes, Dr. Gram, of Copenhagen, my esteemed friend and co-worker, has lately succeeded in discovering a process for the *isolated* staining of bacteria,[*] which, for most cases, may be regarded not only as the best, but as a wellnigh ideal, method for effecting this result.

Oftentimes, indeed, the strongly stained nuclei conceal the bacteria, which are always much smaller, and, consequently, often only slightly colored. For this reason, when the old Weigert process is employed for substances rich in nuclei,—as the spleen, lymphatic glands, the lungs in pneumonia, granulation tissue, etc.,—it becomes necessary to make use of especially thin sections ; and, even then, one often experiences considerable difficulty in seeing and demonstrating these minute organisms.

---

[*] Gram : Fortschritt d. Med., 1884, S. 185.

By Gram's method the bacteria alone remain stained : all else becomes completely colorless ; while the intense dark-blue color of the microbes must render almost every individual organism which is present in the section immediately apparent to the observer.

In the first place the method is purely empirical. Ehrlich's solution of gentian-violet in aniline water is employed for the staining. Finely pulverized gentian-violet is placed in a glass bottle to the depth of about two finger-breadths ; and upon this, aniline water is poured. This aniline water is a saturated solution of aniline (a yellowish, oleaginous fluid) in water, and may be quickly prepared by shaking up 4 parts of aniline in 100 parts of distilled water. The surplus aniline, which partly settles to the bottom of the vessel and is partly suspended in the water, thereby causing a slight coloration, is removed by filtration. This perfectly clear aniline water is, as we have already stated, poured upon the dry gentian-violet ; when, if the mixture be frequently shaken, a saturated staining fluid is quickly (at all events in 24 hours) produced. Almost the same result is obtained by pouring 5 parts of a saturated alcoholic solution of gentian-violet into 100 parts of aniline water.

By loosely closing the bottle with a funnel, armed with filter-paper, any quantity of the solution which may be desired can always be filtered into a small dish without further trouble. The same filter may be employed many times ; and the fluid remains unchanged for weeks and months.

The section of an alcohol preparation (or the cover-glass with its adherent dry specimen) which is to be examined, is placed in this fluid for a few minutes, after having previously lain in absolute alcohol *only*, not in water or diluted alcohol. Now very deeply stained, the section is transferred to a weak solution of iodine and the iodide of potassium (iodine 1 ; iodide of potassium 2 ; distilled water 300), when a dark precipitation ensues. After being in this fluid for 1–3 minutes, the section is placed in absolute alcohol. In this latter medium the specimen surrenders nearly all its staining matter, and the alcohol assumes a purple shade ; while the section is soon as good as colorless. After being rendered transparent in oil of cloves, the section is mounted in xylol-Canada-balsam, in glycerine, or in Kleb's mixture of glycerine and gelatine.

While, therefore, all the elements of the tissues appear completely colorless (a pale-blue shimmer sometimes remains in the nuclei), the bacteria come to the front as dark-blue, almost lividly stained, granules, rods, filaments, etc.

It is clear that this method exposes the bacteria much more distinctly, and in greater numbers, than did the earlier staining processes ; and it is possible that this mode of procedure will enable us to discover the presumptive bacteria in those infectious diseases, of which the pathogenic excitants are as yet wholly, or partly, unknown (syphilis, petechial typhus, dysentery, etc.).

Supplementary tingeing with Bismarck-brown enables one to produce very handsome double staining ; the bacteria remaining dark blue, while the nuclei assume a shade varying from yellow to brown. Thus the precise location of the bacteria within the tissues is more clearly shown. One sees whether they lie without or within the cells, etc.

It remains to be said that the bacilli of typhoid are left unstained by this process ; they being decolorized as well as the nuclei. In this respect they differ from most of the other forms of bacilli. In many cases of pneumonia, too, the cocci become almost completely decolorized by this iodine process. In the majority of instances, however, bacteria appear in vast numbers, infinitely more numerous than one would believe possible, who had employed the old Weigert process. A comparison of the two methods upon sections of the same series is very instructive. Nevertheless, one must consider that there may be cases, in which the iodine method does not disclose bacteria which can be demonstrated in some other manner.

d. Staining of Tubercle-Bacilli.

The peculiar effect of the aniline dyes upon the *bacilli of tubercle* is a matter of great importance. After a considerable time had been passed in vain endeavor to demonstrate anatomically the morbific agent of tuberculosis,—the statements occasionally appearing regarding the discovery of micrococci, etc., in tubercles being untrustworthy,—it again devolved upon Koch to prove the constant presence of a specific bacillus in the products of tuberculosis by the aid of a peculiar staining process.* And inasmuch as one of the most important discoveries of recent medicine is concerned, I should here like to introduce Koch's original method; although it has been forced into the background, and is seldom used, since the publication of Ehrlich's modification, which took place soon afterwards.

Koch gave the following directions : The section, or dry preparation (see page 78), is placed in the following mixture for 24 hours :

Distilled water	200.00
Concentrated alcoholic solution of methyl-blue .	1.00
Liquor potassæ—10 per cent.	0.20

thus receiving a dark-blue stain, after which it is placed for fifteen minutes in a concentrated aqueous solution of

---

* Koch: Die Aetiologie der Tuberculose. *Berl. klin. Wochenschr.*, 1882, No. 15.

vesuvine.   It is now washed in distilled water until the
blue color has disappeared, leaving a more or less deep-
brown tinge ; when, after being deprived of its water by
the action of alcohol, it is cleared with the oil of cloves,
and examined with the aid of a homogeneous immersion
lens and open condenser.   The nuclei, as well as most
varieties of micrococci, now have a brown tinge, as
though they had been exposed to the action of the
vesuvine alone.   The tubercle-bacilli, on the other hand,
are tinged a deep blue.   Koch attributed this result to
the alkaline reaction of the staining fluid, since the
bacilli were never stained in fluids which were acid or
neutral in reaction.   The neutral solution of another
dye-stuff displaces the first color everywhere, except in
the tubercle-bacilli, these latter retaining the original
blue tinge completely.   The process, therefore, depends
upon a specific reaction, by which the bacilli of tubercle
are alone affected—not other forms of bacteria.   Only
the lepra bacilli deport themselves similarly, according to
Koch—they retaining the blue tinge also ; but these are
tinged too by the simple nuclear staining process, to
which the tubercular bacilli react either not at all, or
only in the slightest degree (see Art. 15).

Ehrlich's modification of Koch's process, which
appeared soon after, rests upon the following principles :
In the first place, he does not employ a solution of

potassa to produce the alkaline reaction of the staining fluid, but *aniline*, an oleaginous fluid of a light-yellow color, which, in a saturated watery solution, will dissolve much more coloring matter than does the dilute potash solution. As decolorizers he employs strong mineral acids. In this he acts upon the theory, *that the tubercle-bacilli, which, according to Koch, remain unstained in neutral or acid staining fluids, while in alkaline solutions of the same coloring agents they become stained, are invested by a sheath or envelope, which is permeable to fluids of an alkaline reaction only.*

But, after the action of the alkaline staining fluid, not only the bacilli are colored, but the nuclei, protoplasm, basement-substance, etc., as well; so that the bacilli themselves remain hidden from view. But, if the preparation be now placed in acid, the color is removed from all the remaining structures, and from any other micro-organisms which may happen to be present, because of the acid's marked affinity for the coloring-matter. *Into the bacilli, however, the acid cannot penetrate; as their presumptive enveloping membranes are impervious to acids. Therefore these bacilli remain the only colored bodies in the otherwise completely discolorized preparation, and are thus rendered markedly conspicuous.*

Whether this hypothesis of a bacillar envelope which is impervious to acids and neutral solutions, though permitting the passage of alkaline fluids, be correct or not, the future must decide. As yet, however, it explains most of the known phenomena, and has led to the discovery of a method by which the bacilli of tubercle, and these alone, are deeply stained in both tissues and fluids.

The correctness of Ehrlich's theory has been disputed by Ziehl, who succeeded in staining the bacilli of tubercle by adding carbolic acid (which substance acts for the most part like an acid, though not possessing an acid reaction) to the aniline dyes. Furthermore, Lichtheim,[*] Giacomi, and others, then declared that simple watery solutions of gentian-violet and fuchsine sufficed to stain these specific microbes of tubercle within a reasonable time, which could be greatly shortened when warmth was employed ; though, to be sure, the staining was not quite so intense as that afforded by the Ehrlich method. Since then, a large number of "new methods" for staining tubercle-bacilli have been published ; but, where reliable, they are all more or less inferior modifications of Ehrlich's process. Inasmuch, then, as the latter is completely reliable, so far as we know, and excelled by none other, we shall limit ourselves to an accurate description of this process, which was accepted forthwith by Koch.[†] We ourselves have never had reason to abandon it. An

---

[*] Fortschr. d. Med., Bd. I.

[†] Mittheilungen aus dem k. Gesundheitsamt, Bd. 2, 1884.

enumeration of the other so-called methods would be very tiresome, and void of any real practical value.

The following principles are to be observed. While most other forms of bacteria are quickly and deeply stained in aqueous solutions of gentian-violet, fuchsine, Bismarck-brown, etc., this is not the case with tubercle-bacilli. Here the *long-continued* action of the staining agent (that is, at an ordinary interior temperature) is needed. Hitherto, attempts to impart a deep stain to tubercular bacilli with Bismarck-brown have generally proved unsuccessful ; and we employ for this purpose gentian-violet or fuchsine. As a solvent, we make use of aniline water, *i. e.* a saturated aqueous solution of aniline. After being stained, the preparation is treated with mineral acids—hydrochloric or nitric—in aqueous or alcoholic solution. Those bacilli which retain their color after treatment with strong mineral acids, are to be regarded as tubercular. This property is quite characteristic of these specific germs ; and we would advise that the diagnosis of tubercular bacilli be deemed uncertain until they have withstood this test. To omit the acid treatment, on the score of inconvenience, is to my mind completely unwarrantable. In properly prepared specimens all other constituent parts of the tissues must be completely decolorized, as well as any other forms of bacteria which may chance to be present ; so that, the structural details being completely wiped out (that is, with the full illumination of an Abbé condenser), the tubercle-bacilli alone occupy the field of view. To render focusing easier, and to disclose the precise location of the bacilli in the tissues, the nuclei also may be stained in the ordinary manner, that color being selected which affords the greatest possible contrast with the tinge of the specific microbes.

Ehrlich's method may be described as follows :

The preparation of the staining fluid, gentian-violet in aniline water, has already been described at length (see section *c*). Instead of gentian-violet, fuchsine or various other aniline dyes (magdala, dahlia, methyl-violet, magenta, etc.) may be selected. It is a matter of indifference whether the dye be added to the aniline water in substance, or in the form of a concentrated alcoholic solution.

Of this staining fluid, which may be kept in stock for weeks (aniline water itself decomposes quite rapidly in the light), from ten to twenty drops or more are filtered into a watch-glass ; and in this the section to be examined, or the cover-glass (see page 78), is placed for about twenty-four hours. Then the deeply stained section, or the cover-glass, is washed in a watch-glass filled with distilled water, after which it is placed in a solution of nitric acid in alcohol (3–100). In a very short time, from three to five minutes at the most, the specimen is sufficiently decolorized ; when it is placed in absolute alcohol, and afterward examined in oil of cloves. The cover-glass preparations can as well be rinsed in water, and examined directly in the same medium.

Should a *double staining* be desired, Bismarck-brown is employed, when the preparations are tinged blue ; methyl-blue, etc., when tinged red. It is to be observed, how-ever, that the nuclear stain must not be very intense ; lest the tubercular bacilli be masked in too great a degree.

The *decolorization* may also be effected with a watery solution of nitric acid ; but, in order to accomplish it in the same time, the solution should be considerably stronger,—about thirty per cent. We prefer the more dilute alcoholic solution ; as this does not attack metallic instruments. In case the decolorization has been in-complete, some of the former color reappears, after the acid has been removed by rinsing the specimen in water ; but this residuum may be removed by again subjecting the specimen to the combined action of the acid and alcohol. The stain of the tubercle-bacilli, however, re-sists the action of the acid mixture for more than fifteen minutes.

We are indebted to Rindfleisch for a real improvement of this method ; viz., *rapid staining* at an elevated tem-perature. The watch-glass containing the staining fluid

and specimens is placed in a **thermostat, which is ad-**
justed at 60°–80° C. (140°–176° Fahr.), **or it is simply**
held over an alcohol or small gas flame, till it vaporizes
freely ; when the tubercular bacilli are stained in the
course of a few minutes. This procedure has quickly
met with universal acceptation, particularly for cover-
glass preparations (sputa) ; but it cannot be so highly
commended for **sections.** The further treatment—decolorization, double staining, etc.,—is **effected in the**
usual manner, and at the ordinary temperature.

When treated in this way, the bacilli of tubercular
processes may be seen with an enlargement of only 300
diameters,—often, indeed, without the aid of immersion
objectives. Nevertheless, the best lenses are requisite, in
difficult cases, **to gain a view of the fine** bacilli, which
may, perhaps, lie nearly hidden by other structures.
Before deciding positively, that the bacilli are *absent* from
a preparation—a decision **which may** result very gravely,
—the **prudent** observer will exhaust **the** best-existing
optical **resources. Otherwise, gross errors would be**
unavoidable ; **as the experience of the past teaches.**
Whoever, **then, wishes to search for** these **germs of**
tuberculosis **for** diagnostical purposes, must necessarily
have at his disposal **an** Abbé illuminating **apparatus,**
and a powerful immersion (preferably **homogeneous**
immersion) objective.

Unfortunately, these preparations are **not lasting, for**

the most part ; but the color of the bacilli usually fades, or quite disappears, within a few months, often sooner.

The tubercular bacilli can also be demonstrated without staining. Baumgarten, by treating the sections with a solution of caustic potash, ascertained their presence in some cases of inoculated tuberculosis in animals, later in the human being ; and this, before he knew of Koch's results.

It is a fact, that the bacilli can be evidenced in many cases of tuberculosis by this simple procedure ; but still we shall necessarily be obliged, in all instances, to employ Koch's staining-process, as modified by Ehrlich. In this way alone, we determine whether we have to do with tubercular bacilli or not ; for, as we have already said, the other varieties of bacilli, excepting solely those of lepra, do not respond in this specific manner to the Koch-Ehrlich method. Besides, it cannot be doubted, that, in many cases where the bacilli are sparingly distributed, their discovery is rendered easier, or possible, by reason of their vivid tinge. In the tubercles of fungous arthritis, for example, we shall seldom get a sight of the isolated bacilli by the use of the potash method ; whereas, even here, almost invariable success attends the staining process. Moreover, in examinations for purposes of diagnosis (sputa, pus, etc.), the staining procedure can never be dispensed with.

The stained bacilli of tuberculosis often exhibit bright, colorless spots of oval shape, which are regarded as spores.

ε. The Comma Bacillus of Asiatic Cholera.

Though the position taken by the eminent mycologist, Koch, in ascribing the existence of that deadly scourge, Asiatic cholera, to the direct or indirect agency of the comma bacillus, has been vigorously assailed, his convincing address at the recent cholera conference in Berlin,* as well as his remarks during the subsequent discussion, seem ample warrant for the translator's insertion of a few words concerning this specific microbe, which comes to us from the delta of the Ganges.†

About one half, or at most two thirds, as long as the tubercle-bacillus, the bacillus of Asiatic cholera is thicker and plumper ; while a characteristic curve, resembling that of the comma (,), accounts for its name. Semicircular and S-shaped forms, suggesting the junction of two individual bacilli, also occur ; while, in artificial cultures of the comma bacillus, long spiral threads, very like the

---

* For the full text of Dr. Koch's address, see the *Berliner klinische Wochenschrift* of Aug. 4th and 11th, and the *Deutsche medicinische Wochenschrift* of Aug. 7, 1884. A translation may be found in the *Lancet* of Aug. 9 and 16, 1884.

† For a complete and recent review of this important subject, the reader is referred to " A Treatise on Asiatic Cholera," edited and prepared by Dr. E. C. Wendt for Wood's Medical Library, 1885.

spirochætæ of relapsing fever, appear, leading Koch to doubt whether the organism should not be classed as a spirillum in its various developmental phases, rather than as a genuine bacillus (see plate).

Inasmuch as several different comma-shaped microbes, morphologically identical with the bacilli of cholera, have been discovered, a mere microscopical examination does not suffice for the positive identification of any one member of this group; but, by cultivating them all in various nutrient media, certain biological distinctions are observed, which supplement the information afforded by the microscope, and enable us thereafter to recognize these individuals. The method of cultivating the comma bacilli is wellnigh identical with that employed in the case of the bacilli of typhoid and tuberculosis.*

It should here be remarked, that all vessels and instruments used in culture experiments are to be purified in the flame of a Bunsen burner or alcohol lamp, where this is practicable; otherwise, by exposing them for about fifteen minutes to a temperature of 160° C. (320° F.) in a drying oven, or to steam at 100° C. (212° F.).†

---

* Complete descriptions of the various methods which are resorted to in the study of bacteria, will be found in the " Mittheilungen aus dem Reichsgesundheitsamt," vol, i., 1881, or in a volume entitled "Die Methoden der Bakterien-Forschung," von Dr. Ferdinand Hueppe, Wiesbaden, 1885.

† All such implements and apparatus as are employed by Koch

While hot, the mouths of test tubes are to be plugged with sterilized cotton.

Both fluid and solid culture media are employed, such as bouillon, milk, boiled potatoes, coagulated blood-serum, agar-agar jelly, and prepared gelatine ; though, for the study of cholera germs, the gelatine is preferred. This medium is prepared as follows : 250 grams (half a pound) of finely chopped, fresh lean beef is mixed with 500 grams (one pint) of distilled water, and allowed to stand in a cool place for about 24 hours. The extract is then strained through gauze into a jar, more distilled water being added to the beef, if necessary, in order to bring the strained fluid up to 400 c. c. The jar is now placed in a metal vessel partly filled with water, which is maintained at the temperature of the body ; and to the beef-water are added 40 grams (600 grains) of pure stick gelatine, 4 grams (60 grains) of dried peptone, and 2 grams (30 grains) of salt. The gelatine dissolves in about half an hour, though this time may be shortened by stirring with a sterilized glass rod. Enough of a saturated aqueous solution of the carbonate of soda is now added to the mixture to impart a slightly alkaline reaction, just enough to turn red litmus-paper blue ; when it is carefully boiled over a water-bath for at least

---

in his bacteriological work, such as hot air and steam sterilizers, incubators, damp-chambers, etc., may be obtained of J. F. Luhme & Co., Berlin, N. W., Friedrichstrasse 100.

an hour, or until all the albumen is coagulated, leaving a clear fluid.   A small quantity of this fluid may now be filtered into a sterilized test tube, for the purpose of again testing the reaction, as well as to ascertain whether the albumen has been sufficiently eliminated.   Should a faint cloudiness ensue when the test tube is heated, it is due to the presence of either albumen or phosphates ; in the latter case it disappears on cooling.   These tests proving satisfactory, the whole of the solution is to be filtered through a double thickness of filter-paper,— an operation which may be hastened by carefully heating the filter-funnel with a Bunsen burner or alcohol lamp, the flame being rapidly circulated about the glass.   A hot-water filter is more convenient.

In this the funnel is of metal, and double-walled, the interspace being filled with water, which is kept at an elevated temperature by allowing the flame of a Bunsen burner to impinge upon a short tubular projection of the outer wall.   The test tubes, of which forty or more, holding about 30 grm. (one ounce) each, are required for the filtrate, are prepared by cleansing them, introducing a plug of cotton-wool in each, and heating in the flame of a Bunsen burner or alcohol lamp, both below and above the cotton corks, till the latter become slightly browned.   Now remove the cotton from a test tube with previously heated forceps, fill to nearly one third the

length of the tube with the gelatine solution, and immediately replace the plug. When the food-gelatine has
thus been exhausted, the test tubes are all placed in a
metal vessel partly filled with water, and exposed to a
boiling temperature for about an hour on each of four
successive days, when the sterilization is complete. If
the mouths of the tubes be covered with thin gutta-percha
tissue, to prevent evaporation through the cotton corks,
and especially if the upper ends of the tubes be also occasionally heated, to insure the continued sterility of
their contents, this culture medium may be preserved
unchanged for a lengthy period.

When employed for plate cultures of the comma bacilli,
the contents of three tubes are liquefied by warming;
and, with a previously heated platinum-wire loop or
point, a drop from the alvine discharges, or a mucous
flocculus from the intestines of a suspected case, is transferred to one of the tubes, the germs being diffused
throughout the food-gelatine by agitating the tube (first
dilution). In a like manner two or three drops are
transferred from the first to the second (second dilution),
and from the second to the third tube (third dilution),
the platinum point being sterilized each time, both before and after using it. The fluid contents of the tubes
are now poured upon three sterilized glass plates, each
about six inches long and four inches wide; and, after

being spread out evenly with a sterilized glass rod, the gelatine is allowed to harden under a bell-glass, a bed of ice or snow hastening this process. The plates are next placed, one above the other on glass benches, in a moist chamber ; and the mixed cultures are allowed to grow for twenty-four hours, after which they are examined with a microscope under a magnifying power of about one hundred diameters.

The colonies of comma bacilli are found to assume a characteristic and definite form in this culture medium. When very young, the colony appears like a small, light-colored spot, which is not circular, as bacterial colonies usually are in gelatine, but its contour is more or less irregularly indented, being rough or jagged in places.

At an early stage of its growth it also assumes a some-what granular appearance, which increases with the size of the colony ; till, finally, the latter resembles a small group of strongly refracting granules, like minute frag-ments of glass.

As the colony grows, the gelatine in its immediate neighborhood liquefies, and it sinks deeper into the sub-stance of the nutrient medium, forming a funnel-shaped depression, in the middle of which the colony appears as a small whitish point.

This sinking of the colonies is better shown by pre-paring a pure culture in a test tube. To effect this a suitable colony is sought out on the gelatine plate with the microscope, using a low power ; and, with a fine platinum needle, previously sterilized, a particle is removed from the colony and transplanted into the sterilized gelatine in a test tube. This done, the tube is immediately closed with its pledget of sterilized cotton. Such a culture grows in the same manner as the colony on the gelatine plate. Here, too, one soon observes a small funnel at the point of inoculation. The gelatine slowly liquefies about this point ; while the growth of the little colony may be plainly traced. Above, a deep depression always remains, which looks, in the partially liquefied gelatine, like a bubble floating over the colony. It almost conveys to one the impression that bacillar vegetation caused not only a liquefaction of the gelatine, but a rapid evaporation, as well, of the fluid thus formed. The liquefaction of the gelatine in the test tube slowly progresses, and is completed after the lapse of about a week. These biological characteristics of the comma bacilli of Asiatic cholera, when cultivated in prepared gelatine, suffice for their recognition.

Agar-agar (Ceylon moss), combined with bouillon and peptone, also serves as a cultivating medium. This agar-agar jelly is not liquefied by the comma bacilli.

When cultivated upon slices of boiled potato, the comma bacilli form a thin, pulpy coating, similar to that produced by the bacilli of glanders, but of a lighter brown color.

The movements and development of these specific microbes may be observed under a high power, when they are cultivated in bouillon. This is accomplished by placing a drop of fresh, and completely sterilized, meat-broth upon the centre of a cover-glass, also sterilized, and inoculating this with a particle taken from one of the plate colonies. After coating the edges of the hollow in a concaved slide with vaseline, the cover is inverted, and carefully placed in position. After twenty-four hours, the pure culture, thus obtained, may be examined with an Abbé condenser and an immersion objective.

Dry specimens may be prepared from such a pure culture by carefully raising the cover-glass, and spreading a little of the bouillon over the dry surface of another cover. The fluid is allowed to evaporate in the air, or under the influence of a gentle heat; when the cover is passed through the smokeless flame of a Bunsen burner or alcohol lamp three times, and stained with a drop of an aqueous solution of fuchsine or methyl-blue. After a few seconds the superfluous color is washed off with distilled water; and the specimen may then be examined in water, or permanently mounted in Canada balsam. In preparing dry specimens from gelatine plate-cultures, the distribution of the bacilli over the surface of the cover is rendered easier, and more uniform, by the addition of a drop of distilled water.

When, in a case of suspected Asiatic cholera, it becomes desirable to search for the comma bacilli promptly, in order to distinguish this disease from cholera nostras, acute arsenical poisoning, etc., a small quantity of mucus from the dejecta is spread on a cover-glass, dried, stained and examined, in the manner described above.

To demonstrate the comma bacilli in the intestinal walls, sections are made after the viscus has been thoroughly hardened in absolute alcohol, and these are best stained in a strong aqueous solution of methyl-blue for twenty-four hours ; though this time may be shortened by warming the staining fluid. The sections are then prepared for examination in the usual manner.

The comma bacilli thrive best at temperatures between 30° and 40° C. (86° and 104° F.). When exposed to much higher or lower temperatures, to the influence of most acid media, or when deprived of air, their growth is retarded. It is a striking fact, however, that this microbe is readily killed by drying. When choleraic material is allowed to dry upon a cover-glass for an hour, or even less, the germs frequently perish ; while twenty-four hours without moisture invariably proves fatal.

16.—The Precious Metals.

a. Silver.

The "silver method," which was introduced by Von Recklinghausen, is of great value in the study of normal histology. The important discovery, that the walls of the lymph and blood capillaries, formerly regarded as homogeneous, are constructed of endothelial cells, was, as we know, rendered possible by its use ; and, even at the present day, the method is well-nigh indispensable for the demonstration of this fact. For our purposes, however, the silver process is but seldom employed. One of its simplest offices is to enable us to determine whether a given surface has an endothelial investment ; but the task becomes more difficult, when pathological changes in the walls of blood or lymph capillaries are to be studied.

The difficulties of the method reside in the fact, that the silver salt which is employed—generally the nitrate —forms granular and filiform coagulates with the albuminous fluids of the body, which are of very irregular form, and may easily assume illusory appearances. We desire to confine the precipitation of the silver to the intercellular connective substance, and, accordingly, are limited almost exclusively to normal surfaces in the employment of this method. The action of the silver salt affects the tissues to a very slight depth only.

It is best to employ very dilute solutions,—1 to 500. If necessary, the surface is washed with distilled water, or with a dilute (2-per-cent.) solution of sodium nitrate ; after which the silver solution is poured over it, and, after the lapse of about a minute, it is again washed with distilled water. In a short time, particularly when exposed to sun-light, deep-black lines appear at the edges of the endothelial cells ; while the nuclei usually remain unmarked. These latter structures may be subsequently stained with hæmatoxyline, etc. The silver precipitate is easily dissolved with dilute ammonia.

In order to define the endothelial cells of the blood-capillaries, the silver salt is injected into the arteries ; while, if the solution be forced into the bronchial ramifications, the boundaries of the epithelial cells, which line the pulmonary alveoli, are tinged. By mixing about five per cent. of gelatine with the silver solution—a process which is expedited by warmth,—a very useful injecting-fluid is obtained, which imparts a brown stain to the borders of cells lining the injected cavities.

If a cornea be placed in a solution of silver for a short time, or if a stick of the nitrate of silver be passed over its surface, its proper substance assumes a dark-brown color, within which the corneal corpuscles appear as light, radiate figures, resembling vacant spaces.

At all events, it should be observed that the silver

method of staining is practicable only while the preparations are in a fresh state, or before the supervention of cadaveric decomposition.

b. Gold.

Similar difficulties also beset the chloride-of-gold method of staining, which was introduced by Cohnheim. This agent is possessed of only slight powers of penetration, and has, besides, the fault of inconstancy in its effects. The conditions which govern the reduction of the gold salt, and hence influence the staining, are not yet accurately known. Preparations which have been successfully treated with gold are of great value, by reason of their extremely precise markings ; and the method cannot be dispensed with, particularly in experimental work upon the cornea, keratitis, reproduction, etc.

The advantages of the gold method lie in the following effects :

1. The protoplasm of cells, particularly those of the cornea, receives an intense, dark stain, and is thus sharply distinguished from the completely clear matrix.

2. The axis-cylinders of nerve fibres receive a special tinge, which serves to distinguish them.

Solutions of the chloride of gold, ranging in strength from one tenth to one per cent., are employed. The cornea or other lamellar structures are allowed to remain in the gold solution for from ten minutes to an hour, when they assume a yellowish straw color. They are next placed for some time, about twenty-four hours, in a dilute acid—acetic, formic, tartaric, or citric ; and, for this purpose, Ranvier has lately employed lemon-juice. By this time the reduction has taken place, or it may not be fully completed for several days, during which time the preparation is preserved in alcohol, glycerine, etc. The color is now a dark violet. The gold salt imparts a firm consistence to the specimen ; though, when fine sections are desired, this may be further hardened in alcohol.

Gold chloride can also be used to stain sections of the nervous tissue, which are cut after the usual treatment with Müller's fluid (Leber). Such sections are placed in a one-half-per-cent. solution for about an hour, and subsequently in distilled water. After one or two days, those parts representing normal nerve-medulla, receive a dark-violet tinge. The method is accordingly very useful for the demonstration of degeneration and atrophy in peripheral nerves, and in the white substance of the cerebro-spinal axis.

The following method of staining the nerves with gold, in alcohol preparations, is recommended by Fritsch :

The sections are washed in water, placed in a 0.6 per cent. chloride-of-sodium solution for twenty-four hours, and subsequently in ten-per-cent. formic acid for ten minutes. They are next well washed for from one half to three hours in a one-per-cent. solution of the chloride of gold and sodium, being protected from the light during this period ; after which they are again cleansed in water, and placed in a ten-per-cent. solution of formic acid for twenty-four hours.

c. Osmic Acid.

This reagent, which was first employed by Max Schultze, is now very extensively used, serving :

1. To fix and harden delicate tissue-elements in nearly their natural form.

2. To bring out, or stain, the fats, including the medulla of nerves.

Osmic acid, in solution, also penetrates only the superficial strata of preparations.

Solutions of the acid ranging in strength from one tenth to one per cent. are used ; and it should be stated, that the osmic fumes cause a violent irritation of the conjunctiva, and nasal mucous membrane.

The osmic solution, as well as those of the chloride of gold and nitrate of silver, are kept in brown bottles.

Small pieces of fresh tissue, when placed in a dilute solution of osmic acid for about an hour, and subsequently preserved in glycerine, often afford very good solations of the cellular and fibrous elements, which, by the agency of the acid, have become possessed of a certain power of resistance, in addition to their light-brown stain. This method is especially adapted to nervous tissue, giving to the nerve-medulla a color varying from dark blue to black. The red blood-globules, too, are browned by osmic acid, and rendered capable of marked resistance to most influences. Even the vaporous emanations arising from osmic acid at an ordinary temperature, have this effect ; and it suffices to invert a slide over a bottle containing this agent, in such a manner, that the specimen, with its suspending drop of fluid, may be exposed to the fumes.

When exposed to the stronger and more protracted action of osmic acid, small pieces of tissue, nerves, etc., are hardened.

The various fats, as well as the medulla of nerves, are stained an intense dark-blue color in a few minutes by osmic acid, because of the reduction of the metal, as is generally asserted ; though it is probable, that a special combination takes place. This eminently striking staining-effect, the best which can be secured in the case of fats, is of great value for our purposes. The reaction is also very effective for sections cut from alcoholic preparations ; where that part of the fat which is not dissolved by the alcohol, is tinged a dark brown in less than a quarter of an hour.

To prepare permanent specimens of fatty degeneration in the kidney, liver, heart, granulation-tissue, tumors, etc., for purposes of demonstration, the osmium method can be strongly recommended. The preparations are to be preserved in glycerine.

17.—Sulphide of Ammonium.

Siderosis.

Ammonium sulphide, in aqueous solution, has been very extensively employed by Quincke,[*] in the investigation of pathological tissues, as a reagent for the detection of iron. The iron, contained within the cells as an albuminate, is precipitated in the form of dark-green granules (iron sulphide). The cells which contain iron may often be recognized, even before the action of the reagent, by their yellowish-brown color, but not necessarily so; since all yellow pigment-granules do not assume this dark-green shade with ammonium sulphide. The agent formerly employed for the micro-chemical demonstration of iron, *i. e.* ferro-cyanide of potassium with hydrochloric acid, is less favorable, from the fact that it coagulates albuminous substances, and, moreover, readily gives rise to diffuse stains. The Prussian blue, which is formed, is not wholly insoluble in the albuminous, acid fluid. The dark-green tinge of those granules con-

---

* Quincke, über Siderosis. *D. Arch. f. klin. Med.*, Bd. 25.

taining iron, when ammonium sulphide is employed, appears in sections of alcohol preparations after a few minutes, and lasts for weeks.

Normal red blood-globules do not exhibit this reaction; whence it is to be inferred, that iron is not precipitated from every combination by the sulphide of ammonium. On the other hand, Quincke asserts that in the liver, but particularly in the spleen and medullary substance of bones, " siderosis " is present even in the normal state of these organs ; that is, granules containing iron, which may be demonstrated by the use of ammonium sulphide, occur, whose origin must be ascribed to disintegrated red blood-globules. As a result of transfusion, or so-called artificial plethora, the disintegration of red blood-globules is considerably increased ; and, in consequence, the physiological siderosis is also markedly heightened. In the liver, the iron granules are contained in the white blood-corpuscles of the capillaries ; in the spleen and medulla of bone, they lie in the pulp cells. A very marked condition of siderosis is likewise present in the analogous conditions of the human body, *i. e.* in cases in which an abundant disintegration of red blood-globules takes place. Particularly is this true of pernicious anæmia, in which condition iron may be demonstrated in the hepatic cells and capillaries, as well as in the perivascular connective tissue of the liver ; in the glandular cells of the pancreas ; the epithelial cells of individual convoluted urinary tubules ; and, besides, in the spleen, and medulla ossium.

IV.—OTHER PREPARATORY METHODS.

PRESERVATION OF SPECIMENS.

A LARGE share of the methods of preparation have already been discussed in the preceding chapter.   Hardening with alcohol and the chromic-acid salts, and decalcification by mineral acids, have been considered under their respective headings ; and it only remains to give a short description of a few special processes.

1.—BOILING.

For a long time anatomical preparations have occasionally been boiled, as an aid in histological investigations ; yet, it was only after the suggestions of the departed Perls, that Posner first employed the method in a systematic manner,* principally for the purpose of precipitating the fluid albumen quickly and surely *in loco*, and thus rendering it visible.

Pieces of tissue, ranging in size from a hazel-nut to a walnut, are thrown into boiling water ; whence, after a few minutes, they are removed, and rinsed in cold water.

---

* Posner : *Virch. Arch.*, Bd. 79, S. 311.

They have now usually assumed a moderately firm, elastic consistency, and may be cut with a razor directly; or the hardening may be completed in alcohol.

The coagulated albumen immediately appears in such preparations as a coarsely granular mass ; while most cell-outlines have become strikingly clear, and sharply defined.

The method offers special advantages in the examination of the kidneys, in cases of albuminuria ; and of the lungs, when œdema is present. It is true, that, after simple hardening in alcohol, the granularly coagulated albumen may be demonstrated in the Malpighian capsules or alveoli, respectively, of these organs, and particularly in the superficial portions of the preparations, which were most directly exposed to the action of the alcohol ; yet, this is effected much more completely by the cooking process, which produces a prompt and thorough coagulation in the shortest time. Aside from this effect, however, most structures are only slightly altered by the transient exposure to a boiling heat.

The boiled preparations can also be cut with the freezing microtome.

2.—DRYING.

Specimens were formerly dried, to give them a consistence suitable for cutting. Since the general introduction of alcoholic hardening, however, which, in the case

of porous or very soft objects, is preceded by the imbibi-
tion of mucilage, drying is rarely resorted to, because of
the marked shrinkage which objects undergo, only to be
followed by an irregular expansion, when placed in water.

In the maceration or isolation of certain tissue-con-
stituents,

3.—ARTIFICIAL DIGESTION

plays a serviceable rôle, but is employed more in
normal histology than in pathological investigations.
For this purpose, pepsine or thrypsine, the so-called
artificial gastric juice or pancreatic extract, is employed.

Artificial gastric juice is best prepared from the fundal
portion of the mucous membrane of the pig's stomach.
This is cut into small pieces, and digested in very dilute
hydrochloric acid, 1-1,000, in an incubator at the temperature of the body, after which it is filtered.   The pepsine
of the shops may also be employed ; though its digestive
action should always be tested on a small quantity of
fibrine or loosely coagulated albumen, which should be
quickly dissolved.

The *pancreatic extract* is prepared as follows : *

The pancreas of a freshly killed beef is cut into small
pieces and completely exhausted, with alcohol and ether,
in an extracting apparatus.   The white, friable mass,

---

* Kuehne, in Verhandl. des medic.-naturf.   Vereins zu Heidelberg,
I. 1877.

left after the evaporation of the ether, is digested with five to ten parts, by weight, of a 0.1-per-cent. solution of salicylic acid, or with distilled water at a temperature of 40°C. (104° F.), for about four hours, when it is strained and filtered.

The artificial gastric juice digests connective tissue, muscular substance, most cellular elements, etc., in a short time at the temperature of the body ; while the elastic tissue and nerve fibres resist its action.

The pancreatic extract, or its thrypsine, on the other hand, dissolves in an acid solution the elastic fibres, and, likewise, the delicate fibrils of the neuroglia ; but the connective-tissue fibrils remain intact.

The digestion can either be performed in an incubator, or, by means of the warmable stage, on the microscopical preparation itself, under the eye and continuous control of the observer.

By this method it has been lately discovered that the gray, finely fibrillated substance, which appears in large amount in the posterior columns of the cord in cases of tabes, really corresponds completely, in a chemical point of view also, with the fibres of the neuroglia. It, too, is quickly dissolved ; while the fibres of the pia mater, and its prolongations, remain unaffected. (Waldstein and Weber.[*])

---

[*] Waldstein et Weber, *Arch. de Physiol. norm. et pathol.*, II. Reihe, Bd. 10, 1882, S. 1.

The so-called "neurokeratine" of Kuehne, resists the digestive powers of thrypsine completely. When nerve fibres are digested with hot ether or chloroform, and subsequently with thrypsine, this substance remains behind in the nerve fibres in the form of fine reticula. These "horn-nets" were regarded as preformed structures, a "horny framework of the nerve fibres," by Kuehne and many of his successors ; while Hesse and Pertik, and, most recently, Waldstein and Weber, pupils of Ranvier, declare themselves opposed to this view.

These latter authors assert that the neurokeratine is generally diffused throughout the nerve-medulla, and only assumes the peculiar reticulated structure after the extractive processes have been employed. The form of the reticulum, too, can be changed by varying the method employed. These same "horn-nets," which appear in the interior of nerve fibres, may also be shown in the irregularly formed, extravasated drops of nerve-medulla (myeline), by the proper treatment.

In degenerated nerves, and in gray degeneration of the white substance of the cerebro-spinal axis, the neurokeratine is lost.

4.—IMBEDDING.

(*Celloidine.*)

Most hardened preparations are cut without imbedding. In speaking of the microtome, we referred to the practice now in vogue, of cementing the specimen to cork, or of clamping it between two small pieces of hardened liver.

By these procedures, we are enabled, in most cases without further imbedding, to fix the specimens so firmly, that complete and uniform sections of the same may be made.

In case an irregular surface is to be cut, and it becomes important to include this in the sections, as, for example, in the examination of the endometrium of menstruation, a thin layer of mucilage is spread over the superficies of the hardened preparation, and upon this is placed a flat piece of hardened liver. In alcohol the mucilage soon hardens, and the liver becomes firmly attached to the surface of the specimen. Should the unevenness of the surface be more marked, the mucilage is less advantageous, from the fact that, in thicker layers, it becomes hard as stone and injures the knife. In such instances, therefore, we follow the example of Klebs, and employ a mixture of glycerine and gelatine, which is prepared as follows :

Ten grammes of the finest, well-washed gelatine is allowed to swell up and become soft in distilled water. The surplus water is now poured off, and the swollen gelatine is dissolved by gentle heat (preferably on a water-bath) ; after which ten grammes of glycerine is added, together with a few drops of carbolic acid, the latter for the sake of preserving the mixture from moulding. This mixture

stiffens at an ordinary room-temperature ; but, by warming it, an amount may be melted sufficient to cover the
uneven surface of the preparation. In alcohol it assumes
a consistency proper for cutting.

In other cases we are obliged to cut a preparation *in
toto : i. e.* to make sections which shall include all the
surfaces,—a task which often falls to the lot of the zoölo-
gist and embryologist, but to us only in special cases.
When this occurs, it becomes necessary to invest the
preparation with a mass which hardens on cooling.

For this purpose the glycerine-and-gelatine mixture
may be employed ; though it is better to use a substance
which does not shrink while hardening, or when acted
upon by alcohol. For example, a mixture of wax and
oil, in equal parts ; or, two parts of spermaceti with one
part of the oil of bergamot ; or, again,

Paraffine .	5 parts.
Spermaceti	2 ‘
Lard	1 part.

When moderately warmed, the mixture liquefies, and is
poured into a thin metal box, or mould, adapted to the
clamp of the microtome, and containing the preparation,
which has already been hardened in alcohol. When cool,
the mass is placed in alcohol to harden.*

---

* Instead of the metal box, one of paper, and properly shaped, may
be employed.—S. Y. H.

If we wish the fatty mass to penetrate the specimen itself, the latter must naturally be completely deprived of its water ; and this is usually effected by first permitting it to soak in some ethereal oil, as oil of bergamot. In this way a most uniform consistency is imparted to the preparation, the fat being subsequently extracted from the sections. As we see, this procedure is somewhat long and tedious ; but, for our purposes, it will have to be employed extremely seldom.

Calberla's mixture is also a favorite for imbedding, and consists of

White of egg . . . . 15 parts.

Ten-per-cent. solution of soda carbonate, 1 part.
To this is added the yolks of the eggs used, and the mixture is shaken.*

The object is placed in a paper box, and the latter in a shallow dish filled with eighty per cent. alcohol. The dish is heated up to about 75° C. (167° F.) on a water-bath for half an hour, or until the mass is sufficiently coagulated, when it is hardened in alcohol.

As the imbedding masses are not transparent, their surfaces must be marked, to indicate the position of the specimen.†

---

* Calberla : *Morpholog. Jahrbuch*, II. 1876.

† A pin, passed through the object and into the bottom of the inclosing box, serves to fix the specimen in any position while the imbedding mass is poured in, as well as to mark its position.—S.Y.H.

Perhaps the best imbedding material is " Celloidine," for the introduction of which into microscopical technology we are indebted to Schieferdecker (*Arch. f. Anat.*, 1882).

Celloidine is a collodion-like substance, which is sold in solid sheets, and gradually dissolves to a syrupy consistence in a mixture of alcohol and ether.

If an alcohol preparation be placed in this celloidine solution, it becomes thoroughly saturated in the course of several hours,—it may be twenty-four.

It is next placed in seventy to eighty per cent. alcohol, where the celloidine again hardens, and imparts to the preparation a very uniform consistency, which fits it for cutting. When placed in the microtome, the specimen is incased with a nearly transparent layer of celloidine ; and, accordingly, every section is surrounded by a zone of this substance, in common with which it may be stained, examined, and preserved. To clear up such a section the oil of cloves is not employed, as it dissolves the celloidine, but rather the oil of origanum, or of cedar.

The value of the celloidine method has been demonstrated in many cases, and it may be highly recommended, especially for the eye, spinal cord, etc.

5.—METHOD OF INJECTING.

This artificial filling of the vascular systems with colored masses, or other easily recognizable preparations, is much more frequently resorted to in the study of normal histology, than in the investigations of pathological anatomy.

The study of the *natural injection* of the blood-vessels with blood, of the lymphatic-channels with lymph, etc., is of the greatest importance for us, and we seldom have recourse to an artificial filling of the lumina vasorum. Accordingly we shall give only a short sketch of the methods of injecting, which are quite complicated, contenting ourselves with referring those interested to the admirable text-books of Ranvier and Frey.

a. Injection Fluids.

For purposes of injection we make use of a transparent, though deeply colored fluid, which either remains in a fluid state, or solidifies in the vessels.

In using the latter variety, which usually contains gelatine, both the mass and the organ to be injected must be raised to a temperature of 40°–50° C. (104°–122° F.).

For this purpose, we employ tin chests of considerable size, which are filled with warm water, the temperature of which is maintained by means of an alcohol or gas flame.

The gelatine injections are, therefore, somewhat more troublesome, but, on account of the stability of the mass, they are to be preferred to those with aqueous solutions.

To color the injecting fluid, soluble Prussian blue,

which may be obtained of the druggists, is employed ;
though it sometimes becomes necessary to add to this
some oxalic acid in order to increase its solubility. Such
a solution in from ten to twenty parts of water, can be
directly employed as an injection fluid ; or one may add
five parts each of alcohol and glycerine. Again, the
warmed aqueous solution of Prussian blue may be slowly
mixed with the same quantity of a concentrated gelatine
solution, which is also heated, the fluid being stirred continuously.

The gelatine solution is prepared by putting well-
washed tablets of fine gelatine in about double the quantity of distilled water, and permitting it to soak from one
to two hours at an ordinary temperature. The swollen
gelatinous mass is then dissolved by gently heating over a
water-bath.

Inasmuch as the " soluble Prussian blue " of the druggists is not always reliable, the following most excellent
recipe for its preparation is taken from Frey.*

Thiersch's Prussian Blue with Oxalic Acid.

Prepare a cold, saturated solution of the sulphate of
the protoxide of iron (*A*) ; a similar one of the ferro-
cyanide of potassium, that is, prussiate of potash (*B*) ;
and, thirdly, a saturated solution of oxalic acid (*C*).

---

* " Das Mikroskop und die mikroskopische Technik."

Finally, a warm, concentrated solution (2 to 1) of fine gelatine is necessary. About half an ounce (15 grm.) of the
gelatine solution is to be mixed in a porcelain dish with 6
c. c. of the solution *A*. In a second, larger dish, one ounce
(30 grm.) of the gelatine solution is to be combined with 12
c. c. of the solution *B*, to which 12 c. c. of the oxalic-acid
solution *C* is afterwards added.

When the mixtures in both dishes have cooled to about
25°–32° C. (77°–89.6° F.), the contents of the first dish are
to be added, drop by drop and with constant stirring, to
the mixture in the latter. After complete precipitation,
the deep-blue mixture which is formed is to be heated
up to 70°–100° C. (158°–212° F.) for a time, and constantly stirred, when, finally, it is to be filtered through
flannel.

Preparations injected with the Prussian blue are beautifully stained ; but the color is gradually lost, in the course
of years, by reduction. The blue color is restored by
substances containing ozone, as oil of turpentine.

Carmine injections, on the other hand, are permanent ;
but the use of an ammoniacal solution of carmine is attended with the difficulty, that the red color immediately
transudes in all directions. The alkaline solution must,
therefore, be neutralized, but with the utmost caution ;
else the mass becomes opaque, and is rendered useless by
the coarse carmine precipitate which forms.

Cold-Flowing Carmine Injection.

One gramme of carmine is dissolved in an equal quantity of concentrated ammonia, together with a little water, and to this 20 c. c. of glycerine is added.

To this solution, a mixture, consisting of 20 c. c. of glycerine and 1 c. c. of strong muriatic acid, is carefully added, while the former is being vigorously shaken ; after which the whole is diluted with 40 c. c. of water (Kollmann).

Frey's Carmine Fluid.

Have ready a solution of ammonia and one of acetic acid, of which the number of drops necessary to neutralize each other has been previously determined.

Take thirty to forty grains of the finest carmine, a determined number of drops of the solution of ammonia (the quantity may be greater or smaller, as preferred), and about half an ounce of distilled water ; place them in a mortar, triturate till the carmine is dissolved, and then filter the solution.

These processes require several hours ; and a considerable loss of ammonia ensues in consequence of evaporation.

The ammoniacal solution of carmine is to be mixed with a filtered, moderately heated, concentrated solution of fine gelatine, while stirring. The whole is then to be slightly heated on the water-bath, and the number of drops of the acetic-acid solution necessary to neutralize the original solution of ammonia is to be slowly added, still constantly stirring the mixture. By this procedure a precipitation of the carmine in an acid solution of gelatine is obtained.

If it be intended to inject organs of a more strongly alkaline reaction—those of human bodies which have been dead for some time, for instance,—the acidity of the fluid may be increased by the further addition of several drops of acetic acid. During the injection the temperature should not be raised above $45°$ C. ($113°$ F.).

Recently, other fluids have been employed, especially for the injection of the finest lymph-channels, such as, for example, oily liquids, which are colored by henna ; commonly the oil of turpentine ; or chloroform, in which a dark resinous body, or asphalt, is dissolved.

b. Injecting Apparatus.

. An injecting syringe is often employed. This should be in perfect working order, and, after use, should be very carefully cleansed. The syringe is attached to the canula, either directly or with a piece of rubber tubing ; and both may be of metal or glass. The glass canules can easily be given any desired shape. The piston must not fit the barrel too tightly, and should work smoothly and without any jerking motion.

Though a more complicated process, the injection by means of a *constant pressure*, is worthy of special commendation. If one has not the large machine of Hering at his disposal, he can easily improvise the necessary apparatus with the aid of some glass bottles and rubber tubing. The bottles are closed with doubly perforated rubber corks, armed with glass tubes, of which one is short, reaching only through the cork, while the other descends to the bottom of the bottle. One of the bottles, *A*, is nearly filled with the fluid to be injected; its long tube is connected with the injecting canula, while the shorter tube communicates with the corresponding short tube of a second flask, *B*, which acts as an air chamber, and at first contains air alone. The long glass tube of this second flask, *B*, is connected by a long rubber tube with the pressure-bottle, *C*, which is filled with mercury, and which may be raised to any desirable height by means of an adjustable stand, or blocks of wood.

When the mercury is allowed to flow from *C* into *B*, the air in the latter bottle is subjected to a pressure corresponding to the difference in the level of the mercury in the two flasks. This pressure is brought to bear upon the injecting fluid in *A*, which is thus forced into the vascular system to be injected. The force of injection is kept uniform by raising the receptacle *C* as the mercury flows out, so that the difference in the level of the mercury in the two bottles shall remain **approximately unal-**tered.

Instead of mercury, water **may be used ; when the ves-**sel *C* may be placed on a shelf, or other support, **suffi-**ciently high. As a matter of course, the connections must all be air-tight, and the injecting tube and canula filled with the injecting fluid, or, what is more cleanly, with distilled water, in order that the stream of injection-fluid may not be interrupted by air bubbles. It may easily happen during the injection, especially in the case of structures removed from the body without **proper care,** that the fluid escapes from open branches and collaterals. Such vessels must be tied. It is best that the venous lumina should remain unobstructed at first, in order that any blood present may escape ; but, before the comple-tion of the procedure, these too may be ligated.

By carefully selecting the proper arterial branch, a kid-ney, lung, liver, etc., which has already been incised or cut in twain, may be injected successfully, though only partially.*

The injection is discontinued as soon as the color of the organ has become sufficiently deep ; though the in-creasing consistency of the structures also affords an indication as to when the process should be discontinued.

---

* For this purpose Rindfleisch employs thin, elastic catheters, perforated at the end, as injecting canules. These are pushed deep into the substance of the organ.

Immediately after the injection the specimen is placed in alcohol.

Extravasations are prone to occur in structures which have undergone pathological change, but may, in many cases, be prevented by avoiding too high a pressure when injecting.

When the fine lymphatic vessels in the interior of an organ are to be injected, fine puncturing canules are often employed, which are carefully pushed into the tissues to the required depth.   Necessarily, an extravasation of the injecting fluid results ; but in its vicinity the lymphatics are generally found beautifully injected at the expense of little care or time.

The methods of *physiological injection* have been perfected and extended in a remarkable manner in recent years, and particularly by Cohnheim, Heidenhain, Arnold, Thoma, and others.   In many pathological questions, especially in those relating to the kidneys, they have played a very important part.   They come into consideration, however, only in experimental investigations ; hence we shall refrain from discussing them.

6.—Preservation of Specimens.

In order to preserve, for a few days, fresh specimens which are in a chloride-of-sodium solution, it is only necessary to keep them in a *damp* chamber, and this is very easily prepared in the following manner.

A large, flat glass dish is filled with water to the depth of a few millimetres ; and in this a smaller glass plate, mounted on three legs—made, for instance, of small corks, and fastened on with sealing-wax—is placed, which supports the specimens. A bell-glass, the inner side of which is lined with moistened blotting-paper, is inverted over the glass stage.

The space in which the preparations are is thus kept sealed and sufficiently damp. Instead of the glass plate, a wooden or metallic étagère may be employed, upon the shelves of which the preparations rest without coming in contact with each other. An antiseptic may be added to the water in the dish.

Fresh preparations cannot be preserved for a long time in all their delicacy of structure. Cementing them up in a solution of sodium chloride does not do it, as the elements, for the most part, undergo disintegration in a short time ; moreover, the water evaporates in most cases in spite of careful sealing.

Many structures may be preserved in a nearly saturated solution of the acetate of potassium, which is permanent in air, as is known ; though here, too, much of the original delicacy of the contours is lost.

As a general thing, the preservation of sections from specimens hardened in alcohol is what concerns us. All such are sufficiently preserved when lying in glycerine ;

and we have only to fix the cover-glass, so that dust or other impurities collecting on it may be wiped away. To this end the superfluous glycerine is pressed out by the weight of a leaden bullet, which is placed on the cover-glass, and is taken up with blotting-paper. Next, the slide is very carefully cleansed, every trace of glycerine being removed from about the edges of the cover-glass by means of fine linen moistened in alcohol* ; after which a brush, dipped in a hardening cement, is passed around the cover, fixing its edges to the slide.

The cement used may be either Canada balsam in chloroform, a solution of asphalt in linseed oil and turpentine (Brunswick black), mask lac, etc. ; though for a long time I have employed a thick alcoholic solution of good red sealing-wax, as recommended by Ranvier.†

---

* For this purpose the translator prefers blotting-paper, folded into strips of moderate thickness. One of these is bent upon itself transversely ; when an edge of the fold is slightly moistened with water or alcohol, and passed with firm pressure about the edges of the cover.

† Frey speaks in the highest terms of the Brunswick black used by Bourgogne, of Paris, the composition of which is unknown to him. It dries quite rapidly, and one coating suffices. The so-called mask lac dries very rapidly, and was recommended by Schacht as a cement for wet preparations, and also as a coating for specimens mounted in Canada balsam or copal. The variety of lac used by him is designated as "No. 3" at Beseler's lac manufactory in Berlin (Schützen Strasse, No. 66), and Frey regards it as being only second to Bourgogne's cement. He says : "When concentrated, it forms an excel-

Instead of glycerine, Kleb's glycerine-gelatine mixture (see " Imbedding "), which becomes fluid when slightly warmed, may be used for mounting. Or gum-arabic may be added to the glycerine, so that the mounting fluid gradually hardens at the edges of the covering glass.

Farrant's mixture, consisting of equal parts of glycerine, gum-arabic, and a saturated solution of arsenious acid, is worthy of commendation. It is to be used in the same manner as Canada balsam. For most cases, simply the *mucilago acaciæ* suffices, which, drying at the edges, permanently fixes and preserves the inclosed specimen.

Preparations which have been rendered transparent in oil of cloves, etc., are mounted in Canada balsam, as already stated. Here, too, a further cementing of the edges is superfluous.

V.—OBSERVATION OF LIVING TISSUES.

CIRCULATION, INFLAMMATION.

THE study of pathological processes in living tissues with the aid of the microscope, is practicable only to a very limited degree in members of the human race.

---

lent enclosure for four-cornered covering glasses ; when diluted with absolute alcohol, it is also serviceable for round covers, with the use of the turn-table. It is of a deeper, purer black than Bourgogne's cement, which has a more brownish-black appearance."—S. Y. H.

Hüter's instrument for observing the circulation in the mucous membrane of the cheek, and similar regions, affords only very imperfect results. In the case of warm-blooded animals, too, the difficulties are very considerable, though, it is true, they have been surmounted. Stricker and Thoma have constructed very complicated apparatuses, which enable one to observe for hours the circulation, its pathological disturbances, and the phenomena of inflammation in mammals. For this purpose artificial respiration must be instituted, in order to prevent the death otherwise consequent upon the action of curare. Moreover, the transparent structures, which are exposed to be examined under the microscope,—generally the mesentery,—must be constantly maintained at the temperature of the body. All these efforts, however, have so far resulted in very little ; our knowledge has been only very slightly advanced by laborious observations on warm-blooded animals. On the contrary, those phenomena with which we are already acquainted, were, in the first place, observed on a cold-blooded animal—the frog,—and chiefly by Cohnheim. Because of the great importance of these observations, we shall give a brief description of the very simple methods which are to be followed in making them.

In order to observe the circulation of the blood and its disturbances in the frog, it is well to abolish the voluntary motions of the animal with curare. If a granule of this drug, from 0.5–1.0 mm. in diameter, be placed under the skin of a large frog, the animal becomes motionless in about half an hour ; while the vegetative, or organic, processes continue. The small amount of oxygen necessary can be supplied for days through the cutaneous respiration alone.

For the study of the circulation, one of the first three of the following localities in particular may be selected :

I.—THE WEB.

The use of the web has this advantage : it renders mutilation unnecessary for the observance of the vital processes, the separation and fixation of two toes sufficing. This anatomical structure is, consequently, very useful for many observations ; but it is inferior to the other objects, about to be mentioned, in point of transparency. The pigment cells, too, even in the case of specially selected animals, are disturbing factors, as well as the marked contours of the stratified epithelial cells. Moreover, inflammatory processes proper occur only to a limited degree in the tense tissues. By means of the various irritants, either disturbances of circulation alone, —contraction and dilatation of vessels,—or necroses are produced ; inflammatory swellings do not, as a rule, ensue.

2.—THE TONGUE.

The tongue is drawn out of the mouth and stretched over a cork ring, to which it is fixed with fine insect-needles, or porcupine quills, which are then cut off close to the object. Without further preparation, the tongue is generally insufficiently transparent to admit the use of strong objectives ; and, for this reason, a small piece is removed from the upper (in reality the lower, as the frog lies on its back) surface with a pair of fine scissors, as much care as is possible being taken to prevent hemorrhage, by avoiding the visible blood-vessels. In case bleeding should ensue, however, the blood is washed away with a weak solution of salt. When the field is clear, the cover-glass is put in place, any drying is prevented by allowing the chloride-of-sodium solution to trickle over the exposed surfaces, while the remainder of the frog is enveloped in moist blotting-paper.

The frog, together with the ring supporting the tongue, lies upon a glass plate, which is now placed under the microscope. Observations may then be begun, and may be continued, if desired, for hours, and even days. The tension of the tongue should not be so great as to produce vascular engorgement.

Such an object is wholly adapted to the use of powerful objectives.

The migration of the white and red blood-corpuscles

is well shown,—the wound acting as an inflammatory irritant.

3.—THE MESENTERY.

Large male frogs, recognizable from the glandular protuberances on their thumbs, are selected, in order to avoid being inconvenienced by the oviducts and ovaries. The skin of the lower half of the body is incised in the axillary line, any bleeding which may ensue soon ceasing ; when the incision is completed, and the abdominal cavity opened to the extent of ten to twenty mm. With a pair of blunt forceps a loop of the small intestine is now carefully extracted ; when it is stretched over a cork ring, and fastened just as in the case of the tongue. Too much tension should be avoided, as likely to produce engorgement. The mesentery is now covered with a fragment of covering-glass, and forms a splendid object for examination with high powers. A solution of sodium chloride is employed to prevent the mesentery from drying ; while the frog is protected by an envelope of blotting-paper, or other bibulous material, which is moistened with water. Here, too, the cork ring is not cemented to the glass plate, as it is much more convenient to have it loose and movable. The greater the care taken in the preparation of the object, the more all tension, or other mechanical insult, has been avoided, so much greater becomes the interval ere the exquisite phenomena of inflammation—

the arrangement of the white corpuscles along the walls of the capillaries, emigration, etc.—supervene. The substance and blood-vessels of the tongue, as well as of the mesentery, may, naturally, be exposed to the action of any desired irritant or mechanical injury.

The lung and bladder of the frog can, in like manner, be prepared for microscopical study.

4.—THE CORNEA.

The cornea of the frog too, which survives its removal, affords a good object for the observation of pathological processes, and particularly of inflammation. The normal cornea, or one in which inflammation has been excited by cauterization, for example, is carefully exsected and placed on a slide, together with the small drop of aqueous humor which escapes during the operation. When necessary, several radial incisions are made, to permit of the smooth extension of the curved membrane. In such a preparation, the phenomena characterizing the migrating cells, and the fixed corneal cells, may be observed for many hours.

VI.—EXAMINATION OF FLUIDS.

THE microscopical examination of fluids, for medical and pathologico-anatomical purposes, is exceedingly remunerative. A glance into the microscope often enables us to diagnosticate a diseased condition, which before was veiled in obscurity, or wrongly interpreted. This procedure is attended with few technical difficulties. In the first place a small drop of the fluid is deposited on the slide by means of a glass rod, and covered with a covering-glass.

The drop should not be so large as to rise over the edge of the cover, nor so high as to permit the free flotation of the same—points which are self-evident.

The task before us is the examination of the morphotic elements contained in the fluid.

In very many cases these are evident to the naked eye, whether it be in the guise of a diffuse, cloudy opacity, coarser flecks, or granular precipitates ; and one will naturally first subject such objects to a microscopical examination, by collecting them with a small spoon, or a pair of forceps, and employing objectives of gradually increasing power. *The first point, then, is always an accurate macroscopical inspection of the substance to be examined, and this by transmitted as well as by reflected light,*—a rule which, self-evident as it appears, is lost sight of all too often by the beginner.

In case formed elements are present only in very small numbers, that layer of fluid is examined in which they are to be found in the greatest abundance. This is generally the lower or sedimentary stratum ; as, in the majority of cases, the formed elements are of a higher specific gravity than the suspending fluid, the fatty matters alone floating on the surface. Filtering the fluid and collecting the material thus separated is not to be commended ; as, in this way, impurities cannot with certainty be avoided. In other fluids, again, the morphotic elements are so abundant, that a layer of extreme tenuity must be employed, lest the elements lie in several superimposed strata, and mutually conceal each other.

In case the fluids are very thick and pulpy, it becomes necessary to dilute them before the examination is possible ; and for this purpose we employ serum, or, more usually, a 0.75-per-cent. solution of sodium chloride.

Suspended Elements : Their Vital Qualities— Amœboid Movements.

Except in the last instance, then, we find the elements in their natural menstruum, and may rely upon viewing them in the most unaltered condition possible, provided all prejudicial influences are avoided. And, particularly when studying the *vital phenomena* of elements suspended in a fluid, various precautionary measures must be observed.

In the first place, the pressure of the cover-glass comes into consideration. This may become quite considerable, and is due not merely to the weight of the glass, but much rather to the capillary attraction exerted between the two glass surfaces, separated only by a thin layer of fluid.

When necessary, then, the cover-glass is to be supported by placing fragments of cover-glass, etc., underneath.

Furthermore, the evaporation of the fluid must be hindered to avoid *alterations* in its *density.* This evil begins very quickly at the edge of the preparation, and progresses more slowly at its centre, the rapidity of the process being, naturally, inversely proportional to the depth of the fluid and the distance from the edge of the cover-glass. The evaporation can be reduced to a minimum by covering the object with a sort of *damp chamber.* A glass cylinder of considerable diameter and about two to three cm. in height—a section of a lamp-chimney might answer,—is lined with thick layers of wet, bibulous paper, and placed over the preparation, which lies on a wide slide. Its upper end is nearly closed by the tube of the microscope.

Still better is it to examine a hanging drop of the fluid ; and this may be effected by employing slides having cells from one to two mm. in depth. These may be easily prepared by cementing rings or strips, made of glass or block-tin, to a slide ; or the so-called hollowed glass slides may be used. If the edge of the cell be covered with oil, and a small drop of the fluid which is to be examined be placed in the centre of the lower surface of the cover-glass, an air-tight, sealed apartment is easily formed, within which further evaporation does not take place. It is in this way that we study the movements of the bacteria which may be present in a fluid.   If we arrange two tubes so that they communicate with this hollow space, we are enabled to observe, microscopically, the *action of gases* upon the elements contained in the suspended drop (gas-chamber, or gas-slide).

In this manner are to be observed the so-called amœboid movements of protoplasm, the processes of segmentation in living cells, etc. ; though, naturally, all currents are to be avoided in the fluid, lest rotation of the elements should produce apparent changes in their form. The white cells of the blood and lymph, pus and mucus corpuscles, many of the cells met with in exudates, as well as those of tumors, afford material for these most interesting and captivating studies.   In conducting such observations one must always proceed critically, and with the greatest care,—exercising, above all things, great patience.   The movements are almost always very slow at best, even when the warmable stage is employed.

Of these stages, Stricker's model is to be recommended.

The Form of the Elements.

For the most part, however, we deal with elements of a fixed form ; and it devolves upon us simply to study this accurately.   To do this, it of course becomes requisite that we should view the said elements from all sides ; for it is clear, that a circular figure, for example, as seen in the microscope, may pertain to a disk, as well as to a sphere, cylinder, or cone. So, too, an ellipsoid, an oval, or a still more irregularly shaped body, can, under some circumstances, figure in the field of the microscope as a circle Here we are aided, in the first place, by the fine adjustment, which enables us to view the contour of any object in various planes, and thus to gain an idea of its real form. Again, we are assisted by exciting passive movements of the object, causing it to revolve about its various axes ; and this is accomplished in the simplest manner by causing a current in the investing fluid, either by placing a piece of blotting-paper at the edge of the cover-glass, or by exerting slight pressure upon the latter with the needle. By these manipulations the beginner will oftentimes cause the object not only to roll and re-volve, but to move quite out of the field. Soon, how-ever, he attains the requisite delicacy in the gradation of the pressure, etc., and is then enabled to sharply criticise the configuration of the elements from all sides, thus easily acquiring a knowledge of their stereometrical formation.

EXAMINATION OF THE JUICE OF TISSUES, ETC.

When examining fresh organs, it is often of great im-portance to quickly inspect their elements, cells, etc., in an isolated condition ; and, in many cases, this may be effected in an extremely simple manner, by scraping a fresh surface, and submitting the juice thus obtained to a microscopical examination. For this purpose, a freshly cut surface is always prepared, across which the edge of a scalpel is passed with a scraping motion. In this simple and highly convenient manner, the elements of most parenchymatous organs, or at least some of the same, may be isolated ; the amount of pressure employed being adapted to the firmness of the elements, on the one hand, and to the coherence of the same, or of the cement sub-stance, etc., on the other. Naturally, the marked limita-tions of the method must always be borne in mind ; and yet much time can be saved by its adoption, provided only that the isolated elements be sufficiently character-istic ; as, in many cases, it is by no means necessary to examine exact sections of an organ, in order to answer certain questions.

The juice, which is thus obtained, must, in most cases, be diluted, before it is examined with the microscope ; and, for this purpose, a solution of sodium chloride is generally employed.

In the case of many soft tissues, the juice, with the elements suspended in it, may be sucked up into the lumen of a fine glass tube which is forced into the mass. E. Neumann employs this method, particularly in the examination of the lymphoid medulla of bone, etc. At all events, the elements are thus obtained in their natural menstruum.

The elements of soft tissues may be isolated in a simple manner : by gentle teasing with needles. A fragment of such a tissue is placed on a slide, with a drop of the chloride of sodium solution (0.8 per cent.), and is rapidly torn into small pieces ; when the fluid becomes filled with the isolated cellular elements, etc., which are displaced from their former position. These elements, as well as the tissue-shreds themselves, the latter of which have become sufficiently transparent at their edges at least, are then examined.

In the case of fibrous structures, as muscles and nerves, the elements are separated in the direction of their length, by cautious teasing with the needle.

EXAMINATION FOR MICRO-ORGANISMS.

In view of the great and constantly increasing importance, as well as of the peculiarity of the subject, it behooves us to consider separately the *examination of fluids for the detection of microbes*, and particularly of bacteria.

In the first place, it is clear that in such investigations every contamination must be rigidly excluded. From the time the fluid is obtained, the absolute cleanliness of

the utensils employed must be kept in view.   Furthermore, the examination must always be made while the
fluids are *quite fresh ;* as, owing to the universal distribution of their germs, micro-organisms may be developed
in great numbers in the course of a few hours, when exposed to a suitable temperature.

The germs are present on the sides of every vessel, be
it ever so clean ; on every wiping-cloth, and, though in
smaller numbers, in the atmosphere, particularly the
atmosphere of occupied apartments, sick-rooms, labora-
tories, etc.

In order to collect and preserve fluids free from admixture with the accidental germs of these minute organisms,
special precautionary measures are necessary, among
others the heating of all utensils above 212° F. for some
time.   The assumption of a *generatio æquivoca* (spontaneous generation) of bacteria, which has repeatedly appeared of late years, depended in each instance upon the
neglect of one of the necessary precautions.

Accordingly, we always examine the fluid in a perfectly
fresh state, *i. e.* immediately after its evacuation from the
living body, or from the cadaver ; and, in the latter case,
we shall also have to consider the possibility of the post-
mortem origin of the microbes.

In this connection, a method, for the introduction of
which we are indebted to R. Koch, is to be highly com-

mended. The specimen of fluid is taken up and transferred to the slide on a platinum wire, which is cemented
to a glass rod. The wire may be sterilized very easily and
surely by heating it to a white heat immediately before
and after it is used.

The fluid is first examined in a pure state—*i. e.* without the addition of any thing, in which case one is quite
sure that any organisms which may be found belong really
to the fluid itself.

In many instances the organisms are recognized by
their *lively movements ;* though considerable caution must
be exercised in this respect, inasmuch as small bodies,
suspended in fluids, exhibit almost constantly a dancing,
and, under some circumstances, a very lively motion.
This is the *molecular motion* of Brown. An inadequate
idea generally obtains of the energy of these movements,
which are usually caused by evaporation-currents, etc. ;
and, in order to view the phenomenon for one's self, a
drop of water, to which finely pulverized carmine has
been added, is examined with a lens of high power. The
rapidity and apparent spontaneity of the passive movements undergone by the carmine granules, will at first
cause the greatest surprise. Therefore, before one ven-
tures to pass judgment upon the "spontaneous movements " of granules, etc., which he is disposed to regard
as micro-organisms, he is urgently advised to render him-

self perfectly familiar with these molecular movements of Brown.

Should it be believed, however, that the movements which occur in a given case are vital, the truth of the assumption should always be further strengthened by proving that the motion ceases when conditions inconsistent with the life of the organisms are introduced ; as, for example, heating, or treatment with strong acids and alkalies.

Most of the micro-organisms which interest us—principally the yeast and mould fungi (blastomycetes and hyphomycetes), besides the cleft-fungi, or schizomycetes *—offer a marked resistance to the action of these reagents. The spirochætæ, found in the blood of relapsing-fever subjects, are exceptions, however, as they quickly perish in all reagents, even in distilled water. This resisting power of the microbes can be utilized in their diagnostication, inasmuch as protoplasmic bodies, for example, are dissolved by strong acids and alkalies, while the bacteria remain unchanged.

Particularly when the latter lie in masses, or so-called colonies—the zoöglœa form of bacteria,—they are often made all the more evident after treatment with strong acetic acid or liquor potassæ ; from the fact that the cel-

---

* Bacteria form a subdivision of Nägell's *Schizomycetes*, or cleft-fungi.

lular elements, and other granular masses, which may have concealed the colonies before, are rendered completely transparent.

Their arrangement, too, in rows, or the characteristic form of the individual organisms, as rods, ovals, etc., often renders the recognition of micro-organisms possible without further difficulty. Nevertheless, one should endeavor to avoid errors. Small unorganized precipitates may appear like micrococci, fine crystals like bacilli ; even very minute fat-granules may lead a thoughtless observer into errors.

Koch's Method of Staining Dried Preparations.

Under some circumstances it is by no means easy, often, indeed, impossible, by simple observation and the ordinary micro-chemical reactions, to arrive at a positive conclusion regarding the significance of extremely fine granules which may be present in a fluid. For all such difficult cases, and whenever permanent preparations are to be made, the *drying and staining method*, for which we are essentially indebted to R. Koch and P. Ehrlich, is to be employed.

The method is founded upon the following two facts :

1. When a thin layer of fluid is rapidly dried, the forms of the cellular elements and bacteria contained in it remain essentially unaltered.

2. The bacteria are characterized by a great affinity for the *basic aniline dyes*, and may in this manner be distinguished from other granular structures, etc.

Nevertheless, it is to be remembered, that not only bacteria, but other bodies as well—such as the nuclei of cells and fragments of the same, besides certain protoplasmic granules, for example,—exhibit the same affinity for these coloring matters ; so that in employing this process strict criticism is always demanded in estimating the results.   Moreover, it is conceivable that forms of bacteria exist which do not possess this affinity.   All the varieties with which we are thus far acquainted exhibit a very marked capacity for staining ; some forms, however,—the bacilli of tuberculosis, for example,—only under certain definite conditions.

The process is as follows : A small drop of the fluid is spread upon the cover-glass (or slide) in the thinnest possible layer, either with the aid of a needle or the platinum wire ; or by placing a second cover-glass upon the first, and then removing it.   The beginner is prone to commit the error of employing layers of fluid which are too thick ; though, in this respect, the extremely thin stratum of blood, which is used in examinations of that fluid, affords a proper criterion.   The fluid is then dried in the air, and exposed for a few minutes to a temperature of 120° C. (248° F.).   It also suffices to cautiously pass the glass upon which the fluid has dried through a gas flame three times, at about the tempo employed in cutting bread (Koch). One soon becomes sufficiently practised to heat the object properly. This heating process is particularly necessary in the case of fluids containing much albumen, its special object being to transform the albumen into an insoluble modification. But it must not be continued longer than from five to ten minutes, where bacteria are concerned ; lest the capacity of these organisms to receive stains suffers injury.

The heating accomplished,* the preparation is stained.

A drop of a strong solution of gentian-violet, methyl-blue, fuchsine, Bismarck-brown,—in short, of any of the basic aniline dyes, is poured upon the prepared surface of the cover (or slide), where it is allowed to remain from one to several minutes, and is then rinsed off with distilled water. A brown, blue, or red cloud is now perceived upon the glass.

The examination may follow immediately, a drop of distilled water being placed upon the dried and stained residuum ; when the cover-glass is laid (fluid down) upon a slide. Any water which adheres to the upper surface

---

* In the case of non-albuminous fluids, heating may be omitted. Layers of markedly albuminous fluid, on the other hand, when simply dried, and subsequently treated with staining fluids, are prone to swell and partially scale off ; hence, in such cases, one begins by heating or coagulating the albuminous bodies.

of the cover may be removed with a glass pipette, by sucking or blowing through it.

After again drying the under surface of the cover-glass, which may be most quickly accomplished by blowing upon it through a glass tube, the preparation may be lastingly preserved by replacing the water with a drop of a solution of Canada balsam in chloroform.

When a longer time (several minutes) is desirable for staining, the prepared cover-glass may be exposed to the action of the staining fluid in a watch-glass. In such a case, other granular elements are usually deeply stained also ; and, according to Ehrlich, methyl-blue alone offers the advantage of not over-staining, even after acting for hours.

In such a preparation the form of the cellular elements is generally well preserved ; those changes wrought in the shape of some by the spreading of the fluid (comet-like figures) being easily recognized. The nuclei and bacteria are stained with particular intensity, and are thus rendered strikingly prominent.

Lately, P. Ehrlich specially recommends methyl-blue, in a moderately concentrated aqueous solution, for staining bacteria in dry preparations, the fluid acting for half an hour and longer. The dyes mentioned above, however, and particularly the gentian-violet, have afforded the author about the same results as the methyl-blue.

Every one learning this simple method must come to the conclusion, *that it is undoubtedly the best known for demonstrating micro-organisms in fluids.* (Gram's modification of this method may also be employed.—See p. 98.)

In examining such preparations with a powerful immersion lens and open condenser, one instantly recognizes the sharply contoured and deeply stained micro-organisms, from their characteristic forms and groupings ; and soon learns to diagnosticate as such, incidental impurities or precipitates, and to distinguish them from the bacteria. Precipitates are especially troublesome in fluids containing mucus, as the contents of joints, for example ; inasmuch as the mucus likewise receives a stain of moderate intensity when these dyes are employed. And yet, only a little practice enables one, in this instance also, to interpret correctly the irregularly formed granular-filamentous masses.

The color of the bacteria is usually rendered still deeper, by treating the stained preparations for a short time with a dilute solution of iodine and the iodide of potassium (iodine 1.0, potassium iodide 2.0, distilled water 300.0); and by subsequent treatment with alcohol the nuclei may be decolorized, when an isolated tingeing of the bacteria remains (Gram).

Finally, it remains to be said that, when this method is employed, the *tubercle bacilli are never stained.* In this they distinguish themselves from all other known forms of bacteria. Baumgarten has even suggested that this negative quality, the fact that they cannot thus be stained, be utilized for the rapid detection of tubercular bacilli in sputa.

The *juice of tissues* may be examined for bacteria, by staining a dried specimen, as in the case of other fluids.

The platinum wire, after having been heated to incandescence, is passed over a fresh, clean-cut surface of an organ ; when an amount of fluid, sufficient for our purpose, usually adheres to it. This is then transferred directly from the wire to a cover-glass, and is spread over the surface of the latter in the manner already described. This procedure may be highly recommended, as affording the most convenient and quickest demonstration of the micrococci of pneumonia, for example.

Should one desire to make *original investigations*, he will naturally not be content with the simple demonstration of the micro-organisms, but is obliged to study their nature more closely. For this purpose, again, the drying-method is very convenient. One can very easily prepare fifty or more dried specimens from a single fluid, or a single cross-section of an organ, and may preserve them unstained, in a small box, as long as may be desirable. In this manner, a large number of approximately identical cover-glass preparations are obtained, upon which one is enabled to study the effects of the various chemical reagents, dye-stuffs, etc., at his leisure.

It is already apparent, that the effects of all the basic aniline dyes upon the various forms of bacteria are not analogous ; and, so far as is known, the capsulated micrococci of pneumonia vary most in this respect. However, it may be assumed that further investigations will disclose the fact, that other forms of bacteria, as well, respond to certain definite, specific modes of treatment.

I.—BLOOD.

An examination of the blood is, in accordance with the principles already cited, easily made. A small drop of this fluid—taken either from a considerable amount lost through hemorrhage, or from a freshly made needle-puncture—is placed on a slide, in a cleanly manner, and covered. It is of special import that the stratum of blood should be extremely thin, so that the red blood-corpuscles appear in a single layer only. Before making the puncture, the skin must be carefully cleansed and dried ; and the needle should be raised to a white heat immediately before it is used. In spite of these precautions, however, one must be prepared to meet with some impurities, if only epithelial scales.

It is best to wipe off, in a cleanly fashion, the first drop which appears, and then to approach a cover-glass, so that a small portion of the second drop adheres to its surface. The cover, with this minute specimen, which at most should not exceed a small pin-head in size, is gently laid upon a slide, so that the blood is distributed in an extremely thin layer between the two glasses. When the blood is thus spread over a large surface, one immediately discerns the bright, transparent plasma surrounding the discous red blood-corpuscles, and the white cells, which latter, as is known, are normally few in number.

Besides these, small granules of irregular form are found in variable abundance even in normal blood, which are spoken of as elementary granules, or disintegration-bodies (Zerfallskörper). Their significance is still a disputed subject. It is probable that elements of quite dissimilar importance are concealed among these. The small blood-plates (Blutplättchen), recently described by Bizzozero, and supposed by this author to have an intimate bearing upon the coagulation of the blood, have also been hitherto regarded as indifferent bodies, produced by disintegration processes. Hayem styles them "hæmatoblasts," that is, elements from which red blood-corpuscles arise ; though this view is, in all probability, incorrect. These small bodies vary greatly in number and size, and it is not yet definitely known that they possess any pathological significance. Lostorfer and Stricker once regarded them as characteristic of syphilitic blood, and named them " syphilis corpuscles " ; but, were this view correct, every one would certainly be syphilitic.

Distilled water, as well as acids and alkalies, causes the red blood-globules to swell and lose their color, as we know. The hæmoglobine is rapidly extracted, so that these corpuscles vanish from view almost completely. To preserve them as nearly as possible in their natural condition, solutions of salt of a definite concentration (0.75–1.0 per cent., etc.), so-called indifferent fluid media, must be employed. Chloride of sodium solutions of greater strength likewise preserve the blood-corpuscles, but at the expense of their form and size, as they cause them to shrink.

In pathological examinations of the blood, it is in all cases advisable to examine it directly, and without any dilution. The changes in the blood, which are oftenest met with, are the following :

a. Diminution of the Number of Red Blood-Corpuscles in Anæmia.

A little practice enables one to recognize the more marked alterations of this sort directly, and without the aid of a special apparatus, by comparing specimens of pathological and of normal blood which have been prepared in the same way. When accurate determinations are to be made, however, a counting apparatus, or hæmatometer, is employed ; and of these, the instrument devised by Thoma,* and constructed by the optician Zeiss, in Jena, may be specially commended. Exact directions for its use may be found in the paper referred to above. While the number of red blood-globules in a cubic milli- metre of normal blood is nearly five million, in cases of marked anæmia this may sink to 500,000,—yes, to 143,000, in a cubic mm. (Quincke).

It is clear, that such high degrees of anæmia can be plainly recognized under the microscope without a formal counting of the blood corpuscles.†

b. Alteration of the Size and Form of the Red Blood Corpuscles (Poikilocytosis). Nucleated Red Blood Corpuscles.

Normally, as is known, all red blood-globules have the same characteristic disk-like form, with a depression on each side. The centre is thinner, and hence less deeply colored than the border. The size, too, of the normal red globules varies within comparatively narrow limits, a fact of which one may easily convince himself. In most

---

* Lyon und Thoma, über die Methode der Blutkörperzählung, *Virch. Arch.*, Bd. 84, S. 131.

† The hæmatometer of Hayem and Nachet is comparatively sim- ple, and affords quite accurate results. For further particulars con- cerning the construction and use of this instrument, see article by M. G. Hayem, *Gazette Hebdomadaire*, 1875, p. 291 ; also one by Dr. Keyes, of New York, in the *Amer. Jour. of Med. Sciences* for January, 1876.—S. Y. H.

cases of anæmia, but with especial regularity in the so-called essential, pernicious variety, there are, in addition to normal blood-globules, numerous irregularly-formed and, for the most part, smaller bodies containing hæmo-globine—the so-called *microcytes*. Sometimes, also, these cells exceed the normal red blood-globules in size—the *megaloblasts* of Ehrlich.

Nucleated blood-globules, too, have been found, though seldom, in the direct examination of anæmic blood ; and Ehrlich has discovered by the study of stained dry specimens of blood,—prepared by the method already described as Koch's, except that the heating is longer continued,—that nucleated blood-globules can, in this manner, be demonstrated in all marked anæmiæ, whether they be traumatic or essential. At the same time he determined the diagnostic fact, that in traumatic, secondary anæmia, nucleated blood-globules of the size of normal red blood-corpuscles are found—*normoblasts ;* while the large forms—*megaloblasts*—are characteristic of essential anæmia.

Of the *microcytes* and *poikilocytes*, it remains to be said, that they most probably represent altered, or degenerated, red blood-globules. If the blood of a cadaver be examined some twenty-four hours *post mortem*, analogous forms are found in abundance. Under some circumstances they are said to form in a specimen of blood under the eye of the observer (Vulpian). At all events, the blood should be in as fresh a state as possible, when it is examined for these bodies. Some forms of microcytes, however, are certainly to be regarded as factitious.

c. Increase in the Number of White Blood-Corpuscles.— Leucocytosis and Leucæmia.—Diverse Granulation of Protoplasm.

In many diseases, and particularly in febrile conditions, the white blood-corpuscles are increased in number, both absolutely and in proportion to the red (leucocytosis of Virchow).

Normally, the proportion of white to red blood-globules is one to three hundred, or even less. In the direct examination of the blood,—and in this instance, also, it is best to employ the blood in a pure state,—one easily learns to estimate the relative increase of the white globules with approximate certainty.

In *leucocytosis*, which may be a transitory affection, the white corpuscles may be present in the proportion of one to fifty, and higher; while in *leucæmia*, which generally indicates a lasting and necessarily fatal disease, the proportion increases to such an extent in the severest cases, that the white globules exceed the red in number. At the same time, the actual number of red blood-corpuscles is very considerably lessened. For the exact determination of these proportions, a counting apparatus or hæmatometer, is necessary.

The properties of the protoplasmic granules in white blood-corpuscles, lately the subject of **Ehrlich's investigations,**[*] are specially interesting. He dries extremely thin layers of blood, and exposes them for some time to a temperature of about 120°C. (248°F.). By exposing such preparations to the action of various staining fluids, he succeeds in obtaining constant differences in the staining of the granules of protoplasm within the leucocytes, which are of great physiological, as well as diagnostical, value. Ehrlich thus distinguishes between five different kinds of granulations, lettered from *alpha* to *epsilon*.

The α granules, also called *eosinophilous*, are characterized by a marked affinity for acid coloring agents, as eosine. These "eosinophilous" granules are present in very few normal white blood-globules of the human being; and their presence, in large numbers, enables one to distinguish a beginning leucæmia from an ordinary leucocytosis.

The demonstration of the eosinophilous cells is very simple, according to Ehrlich. A specimen of blood, dried in a thin layer on a cover-glass or slide, and

---

[*] Ehrlich's account of the modes of staining employed and their results, is contained in several dissertations of his pupils, and a number of short, scattered notes,—a most unsatisfactory mode of publication.

exposed to the action of heat, is stained in a short time with a drop of a glycerine-eosine solution, and then washed. It is now examined directly ; or may be again dried, and mounted in Canada balsam. If the eosinophilous cells are present in increased numbers, the fact immediately becomes apparent, because of their red tinge.

d. **Other Cellular Elements, Animal Parasites, and Bacteria Occurring in the Blood.**

Cells of considerable size, each of which includes several red blood-corpuscles in its interior, have been found in the blood of typhoid-fever patients (Eichhorst). In all probability, these have their origin in the spleen ; as we regularly meet with an abundance of similar structures, when examining this organ in patients who have succumbed to this disease. Flat cells, too, containing fat-drops in large numbers, have often been discovered in the blood in cases of acute infectious diseases, and of relapsing fever, in particular. These are regarded as vascular endothelial cells.

In severe cases of malarial infection, there are present in the blood, cells, containing black pigment in granules and masses, as well as pigment in a free state—melanæmia.

Tumor elements, which circulate in the blood and provoke metastasis in cases of malignant growths, will hardly be looked for in the examination of capillary blood, as obtained from a needle-puncture, or a cupping-glass. As a general thing, these are characteristic only when of considerable dimensions, and, so, unable to pass through the capillary lumina.

Of the animal parasites, the *filaria sanguinis* and *distoma hæmatobium* occur in human blood ; but only in tropical or sub-tropical regions.

Thus far, bacteria have been regularly found in the blood of the living human being in two diseases only : the Bacillus Anthracis (Davaine) in cases of malignant pustule (charbon), and the Spirochæta Obermeyeri in relapsing fever.* A very thin layer of the undiluted blood is examined ; or dried preparations are made, heated in the manner already described, and colored with gentian-violet, or methyl-blue, etc.

The bacilli of anthrax are slender, motionless rods, which resist the action of most reagents. The spirilla of relapsing fever, on the other hand, are in lively motion, and may easily be destroyed by nearly all indifferent fluid media, and even by distilled water. The occurrence of the spirilla, as is known, is limited to the febrile stage of the disease ; and only seldom are they still present for a short time after the febrile movement abates. But, because of their constant presence during the period of

---

* Tubercle bacilli have, as yet, been found only in the blood of the cadaver by Weichselbaum, the cause of death being general miliary tuberculosis.

fever, they may be regarded as a trustworthy diagnostic criterion for this disease. It is true, however, that sometimes, and that too in fairly severe cases, few of the organisms are present ; and these may easily be overlooked in a hasty examination. In such cases, when a diagnosis is deemed important, it is advisable to abstract several grammes of blood with a cupping-glass, and to allow it to coagulate. The spirilla are wont, under these circumstances, to gradually collect in the borders of the clot, often forming large groups of twenty, or more ; and, may be, rolled up in a ball-like mass, or bound together radiately. Retaining their lively ciliary movements for hours—yes, for days, outside of the body and at the ordinary temperature of a room ; * they excite, when lying in groups, violent current-like movements of the surrounding fluid, and thus draw attention to themselves, even when low magnifying powers are used. For their more careful inspection, an enlargement of three to four hundred diameters is necessary. One can tinge the spirilla in dried preparations of the blood very easily, with the various basic aniline dyes (gentian-violet, etc.).

As yet, the life-history of the spirilla is unknown. Whether the small corpuscles and double granules, in apparent motion, which have been seen in the blood of

---

* It has even been asserted, lately, that they multiply outside of the body (Albrecht).

relapsing fever, or other infectious diseases, and sometimes even in normal blood, and which have been pronounced "micrococci" or spores, may really be regarded
as such, still admits of doubt.

At all events, the statements regarding the occurrence
of micro-organisms in normal blood are by no means
trustworthy. So, too, most of the communications concerning the finding of so-called micrococci, monads, and
bacilli in the blood of patients suffering from various infectious diseases,—such as diphtheria, hospital gangrene,
erysipelas, etc.,—as well as in intermittent fever, are not
sufficiently authentic. Even in pyæmia, and ulcerative
endocarditis, an examination of the blood during life usually discloses no organisms ; while, in these diseases, the
capillaries are found, *post mortem*, choked in many places
with micrococci. In such cases it is permissible to think
that the organisms gained entrance into the blood in
batches, perhaps only a few at a time, but only to be
quickly arrested in their course by the capillaries ; where,
under some circumstances, they might multiply rapidly.

It remains to say, that it is uncertain that micro-organisms have ever, thus far, been found in human white blood-
globules, though often enough in pus corpuscles, etc.; and
yet the former cells are wont to take up other finely granular masses with avidity.

Perchance the future will witness futher disclosures in

this direction. It is quite possible that important structures are sometimes represented by what we have hitherto been forced to designate indiscriminately as " disintegration corpuscles " (Zerfallskörperchen) ; and that we shall some day learn to distinguish between these.

e. Examination of Blood Stains.—Hæmine Crystals.— Hæmatoidine.

Blood which has dried upon wood, linen, metallic instruments, etc., often becomes the object of medico-legal investigation. By softening such stains with suitable fluids—such as an o.8-per-cent. solution of sodium chloride, or a 33-per-cent. solution of liquor potassæ,— blood corpuscles are very often obtained, the form and size of which may frequently be fairly well preserved.

Now, inasmuch as we know that human blood, and that of mammals in general, is characterized by the circular form of the red globules, while, in the case of the other vertebrates, these have an oval shape, it is usually possible to decide positively whether a given blood-stain is of mammalian (including man) origin or not. But our art goes no further. It is impossible for us to diagnose human blood-globules with certainty. It is true that most mammals have somewhat smaller red blood-globules than man, those of the sheep and goat being on an average little more than half as large ; but, on the other hand, some animals—the dog, for example—bear a striking resemblance to the human being in this respect. At all events, in passing judgment upon blood-stains, it is well, in any given case, to limit one's self to the statement that they are, or are not, of mammalian origin.

The size of the red globules does not afford a certain criterion in the determination of their origin ; for, not only does this vary in each kind of blood, but, in the case of blood-stains, differences in the mode of drying and subsequent moistening, in the age of the stains, etc., introduce new elements of uncertainty as to their exact original dimensions.

The so-called *hæmine crystals of Teichmann,* muriate of hæmatine, are also found in dry blood-stains, and are produced as follows :   A small amount of the dried blood—a thread impregnated with it, for example—is placed upon a slide, and to it are added a few drops of glacial acetic acid, together with a granule of sodium chloride.   The slide is then gradually heated till ebullition begins.   Observed now through the microscope, a large number of dark-brown, rhombic crystals are seen to form around the fibre, which are completely insoluble in water and exquisitely doubly refracting.   This test affords positive results even in the case of old blood-stains, and, as may be perceived, is of universal applicability.   It is often necessary to first extract the doubtful stain with water ; when, after evaporation, the extractive may be treated in the above-described manner.

Besides these artificial products, crystals of *hæmatoidine*, also rhombic in form, are found in old blood-extravasations, in the corpora lutea of the ovaries, etc., which are partly free, partly included in cells. They are of a lively ruby, or orange-red color, contain no iron, are soluble in chloroform, and possess the greatest resemblance to bilirubine ; indeed, by many the two are regarded as identical.

According to our present knowledge, crystals of hæmoglobine do not occur in man when in a normal state, but they are found in abundance in the post-partum uterus of the Guinea-pig, among others. They can be produced in a variety of ways, however ; as, for example, by the action of a concentrated solution of pyrogallic acid upon blood, which has previously been diluted with distilled water.*

2.—The Sputum.

The microscopical examination of the sputum is of great diagnostical value, and, therefore, very frequently undertaken. Its technology is very simple.

---

* Solutions of hæmoglobine exhibit a well-marked and peculiar *spectrum*, which is characterized by two absorption bands between the lines " D " and " E,"—the one in the yellow, the other at the commencement of the green. Hence the spectroscope enables us to detect the *presence of blood* in fluids, even after the red globules have been dissolved, and the coloring matter reduced to a fluid condition ; while the aid of the microscope must be invoked to show whether the blood is that of a bird, reptile, or fish, or whether it *may be* of human origin.—S. Y. H.

In the first place, however, this secretion, which always represents a mixture of various substances of diverse origin, is inspected *macroscopically ;* and, inasmuch as the elastic fibres, which are the special objects of our search, are mostly found in opaque white or grayish masses, these latter are carefully looked for, our quest being facilitated by pouring out the sputum in a thin layer upon a black porcelain plate. These elastic fibres, or shreds of elastic tissue, are to be regarded as tissue from the walls of pulmonary cavities. The so-called "asthma crystals" of Leyden, also, generally lie in the interior of greenish-white masses about the size of a millet-seed, which can be distinguished with the naked eye in the otherwise clear sputum. So, too, when there is reason to suspect the presence of echinococci, or other uncommon objects,—in short, in every case—an accurate macroscopical examination should lead the way. A neglect of this rule often enough results in failure and negative results.

Microscopical preparations are next made from all these various parts, which frequently lie in superimposed strata. A portion of the sputum is transferred to a slide with a glass rod, a spatula and needle, or, in case the secretion is very dense and tough, with forceps and scissors, after which the cover-glass is adjusted. An indifferent fluid is rarely necessary for diluting such specimens ; but, if so, use is made of the chloride-of-sodium solution, or of distilled water. When examining with the microscope, one should first employ a low magnifying power,—say fifty to eighty diameters,—exchanging this for a higher only when familiar with the whole preparation. The elastic fibres are usually recognizable, to some extent, even with a lens of low power, either directly, or by the somewhat darker crumbling material in which they are imbedded ; and, inasmuch as the lower powers expose to view a correspondingly larger field, the probability of their discovery is enhanced by the use of such lenses.* Besides, the low power is not so sensitive to slight variations of focal distance, thereby permitting the simultaneous examination of several superimposed strata in a preparation.

Such parts of the preparation as may require it, are next analyzed more closely with a stronger lens, before the microscopical diagnosis is determined upon.

a. The Oral Fluids.

In the examination of sputa, substances of the most diverse nature are almost invariably encountered. One can never expect to obtain the contents of the bronchial ramifications in a pure state, for the secretions of the mouth, pharynx, and salivary glands, at least, are constant

---

* Inasmuch as the beginner might accidentally mistake threads of lint, etc., for these lung-fibres, it would be well for him to consult an experienced observer before committing himself.—S. Y. H.

admixtures ; and, accordingly, one must familiarize himself with the microscopical characteristics of all these constituent fluids.

Normally, *epithelial cells from the mucous membrane of the mouth and pharynx*, generally in a condition of more or less advanced cornification, are always present in the oral fluids in abundance. These large, flat cells of irregular form, which swell up and form spherical vesicles when acted upon by acids or alkalies, one soon learns to recognize. They are generally studded with numerous bacteria. In catarrhal conditions of the mouth and pharynx, living cells also appear in the saliva, which put forth prolongations of various shapes and undergo slight amœboid movements.

The superficial layers of epithelial cells are also frequently exfoliated *in continuo* in oral catarrh ; and, in such individuals, the terminal prolongations of the filiform papillæ, consisting of dense, horny epithelial cells of marked coherence, are especially abundant in the coating of the tongue.

Besides these, we find *round cells* in abundance, *mucous* or *salivary* corpuscles, from the mucous or salivary glands, particularly the submaxillary and sublingual, or, to a less extent, from the deeper layers of the stratified epithelium. Originally, these were small amœboid cells, analogous to the lymph-corpuscles, etc. ; but they undergo changes, when exposed to the thin secretion of the parotid gland,—swelling up into a spherical shape, and disclosing a limiting membrane, which contrasts with the clear contents of the cell. This latter substance harbors one or two roundish nuclei, besides a large number of fine granules ; and the latter are constantly undergoing a lively, dancing movement—*molecular movement in the salivary corpuscles.* These granules are probably not parasitic organisms, as has been supposed ; at least, they are never stained by basic aniline dyes. Whether their active motion represents a vital phenomenon, somewhat analogous to the protoplasmic currents in plant-cells, is not ascertained.

But besides these, the most various *micro-organisms* disport themselves freely in the fluids of the mouth : long, motionless leptothrix filaments, of variable breadth ; spherical cocci of different sizes, arranged often in chains, or compact heaps ; and, not infrequently, bacilli, too, and elegant forms of spirochætæ, which closely resemble the spirilla of relapsing fever in their appearance and serpentine movements, though usually their dimensions greatly exceed those of the latter. The less care employed by an individual in cleansing his mouth, the greater will be the mass of micro-organisms present in that cavity ; though the most painstaking cleanliness never insures their complete absence. The bacterial germs, which are always suspended in the air we breathe, find in the buccal fluids conditions most favorable to their support and propagation. One species of these microbes was regarded as the cause of dental caries, but this view is probably incorrect (W. D. Miller). When introduced into the circulation of animals, many varieties of the schizomycetes found in human saliva act as pathogenical factors.

Of the remaining groups of fungi, the blastomycetes (Sprosspilze) are only occasionally present in any considerable quantity in the oral fluids ; while, on the other hand, of the hyphomycetes (Fadenpilze), the thrush parasite, *oidium albicans,* is frequently met with, and particularly in the case of young sucklings, or in adults whose nutrition is greatly below par, as, for example, the victims of phthisis, etc. This vegetable parasite consists of branched, jointed filaments with oval spores, which form a more or less dense mycelium between the superficial epithelial cells and on the surface itself.

Besides these structures, there occur all kinds of food-remnants, of a vegetable order, in the saliva. Many retain microscopical samples of edibles for many hours after their enjoyment in the oral cavity, and particularly between the teeth, or in carious cavities of the same ; and these often enough excite astonishment and perplexity in the mind of the beginner who is examining the sputum.

The mucous secretion of the nasal cavities, frequently mixed with blood, may also be found in the sputum, and, furthermore, the contents of abscesses opening into the mouth or pharynx.   In this connection, attention must be called to the concretions, ranging from minute particles to masses larger than a pea, which form in the tonsillar crypts, and are due to the calcareous incrustation of masses of retained secretion.   These are occasionally expelled by the act of coughing, etc.; when the patient, and frequently too his physician, is greatly disconcerted by the putative pulmonary calculi.   The diagnosis is easily made, however, with the aid of the microscope.

By the use of dilute hydrochloric acid the lime is dissolved ; when there remain the large, cornified epithelial cells, which form the principal part of the mass, and which are sometimes arranged in concentric layers, together with micro-organisms in prodigious numbers.

b. Products of the Respiratory Mucous Membrane.

Only after an accurate acquaintance has been formed with all these elements of accidental occurrence in the sputum, can one proceed to the microscopical examination of this fluid with advantage.   The principal constituent of the sputum is secreted by the inflamed respiratory mucous membrane,—a mucous fluid, containing round cells in greater or less abundance.   The secretion of the mucous membrane lining the pharynx and the upper part of the larynx, where the epithelium is stratified and of the squamous variety, is, moreover, always very rich in desquamated epithelial cells. On the other hand, an epithelial desquamation occurs but seldom throughout the remainder of the respiratory tract : the lower portion of the larynx, the trachea, and bronchial tubes—*i. e.* wherever the mucous membrane is covered with columnar ciliated epithelium. It is rarely that one meets with ciliated cells, or portions of the same, when examining sputa.

The glassy, transparent portion of the sputum is very poor in formed elements. The more abundant the latter, the more cloudy and opaque becomes this secretion. Round cells are most abundant in purulent sputum ; in which case fine granules of fat usually abound in the cellular protoplasm, increasing the opacity and producing the yellowish tone of the mass.

In the majority of cases the round cells are already lifeless, rigid, and consist of a dark, granular protoplasm, with one or more nuclei. A special parietal layer is usually wanting, the edges of the cells having a finely serrated appearance, as though corroded. The granulation of the protoplasm is, for the most part, so marked as to conceal the nucleus or nuclei. Acetic acid, however, causes most of the protoplasmic granules to disappear ; when the single or multiple nuclei become more apparent.

Most of the round cells which we find in the sputum
are *about the size of the white blood-corpuscles,* or slightly
larger ; indeed, it may well be assumed that a large share
of these, at least, are emigrated white blood-globules,
while others may come from the mucous glands, or the
inflamed tissue of the mucous membrane. Besides these,
however, one not infrequently meets with *larger epithelioid
round-cells* in the sputum, which are characterized by a
sharp, nearly circular contour, and by a vesicle-like nucleus ; *i. e.* the nucleus—seldomer still the two nuclei—is
bounded by a dark, sharply-drawn line, while its interior
is bright, and contains one or more nucleoli.   When
stained—and this is best effected by treating a dried and
heated specimen of sputum with basic aniline dyes, according to Koch's method,—the nuclei of these epithelioid cells do not take such a deep, and uniformly dark,
tinge, as do those of the ordinary, small round-cells ; but
exhibit likewise a darkly stained periphery, and a light
centre.   Some fat-granules, also, are now and then found
in the protoplasm of these larger cells ; often, too, *black
pigment, in the form of minute granules, or particles of
irregular shape,—so-called " lung-pigment."* We often find,
even upon macroscopic inspection of the sputum, blackish or dark-gray spots and streaks ; and the microscope
shows us, that the blackish color is produced solely by
the large number of pigment-bearing cells, free pigment

being usually present in only small quantities. These black granules and particles within the cells are now to be regarded as inhaled coal-dust, which has been deposited upon the mucous surface of the respiratory tract, and there retained, to be subsequently taken up by the amœboid cells.

Such pigmented cells are almost never wanting in the secretions of the mucous membrane lining the pharynx and respiratory tract of individuals, who have ample opportunity to inhale coal-dust; as those engaged in handling coal, hard smokers, and others. Yet, every one who lives as modern civilization decrees—dwelling, as he does, for the most part, in closed and heated rooms, or exposed, even when in the open air, to the emanations of smoking chimneys,—has occasion to introduce finely divided coal-dust into his respiratory tract with every inspiration. The more mucus there is secreted upon the surface of the respiratory passages, the greater will be the amount of coal-dust retained ; and, of this, a portion is absorbed by the amœboid cells and transferred into the pulmonary interstitial tissue and lymphatics, while the balance is again expelled with the sputa.

True pigment, that is, pigment originating in the organism itself, likewise occurs in the sputa, but only very seldom. It is characterized by its brownish (not black) color, and points either to a previous hemorrhage, or to an engorgement of the pulmonary venous system, with migration of red blood-corpuscles and the subsequent pigmentary metamorphosis of the same (brown induration of the lungs). Crystals of hæmatoidine also appear at times.

In the larger, round epithelioid cells of the sputum, already mentioned, a deposition of homogeneous, dully glistening granules sometimes occurs also ; and, because of a purely external resemblance, these are styled "myeline granules." Nothing definite is known regarding their nature and significance.

In the familiar writings of Buhl,* an especial importance is attached to these larger round-cells of the sputa. They are regarded as desquamated epithelial cells from the pulmonary alveoli, and their abundant occurrence in the expectoration suffices for the diagnosis of desquamative pneumonia, or incipient phthisis. These views have proved untenable ; as *large numbers of these cells*, crowded together in heaps even, and containing granules of "myeline," fat, and pigment, are quite frequently found in the *morning sputa of healthy persons*,† as well as in the sputa of those suffering from simple bronchial catarrhs. It is

---

* Buhl : "Lungenentzündung, Schwindsucht, und Tuberculose.' München, 1872.

† Vgl. Guttmann und Schmidt : *Zeitschr. f. klin. Med.*, Bd. 3. Panizza : *Deutsch. Arch. f. klin. Med.*, 1881. Bizzozero : *Centralbl. f. klin. Medicin*, 1881, S. 529.

also highly uncertain that these should really be regarded as desquamated alveolar epithelial cells in all cases. On the contrary, it has been rendered probable by many investigations, that, under some circumstances, ordinary wandering-cells—lymphoid, etc.—may become transformed into such epithelial-like elements.

Among others, cells loaded with fat-granules are sometimes found in subacute pneumonic processes ; and, in tubercular phthisis, giant-cells also, though these latter are a rarity.

c. Elastic Fibres, Fibrinous Effusions, Asthma Crystals.

The discovery of *elastic fibres* in the sputa is, naturally, a matter of very great significance, inasmuch as these show that a destructive process is going on within the lung. As already stated, the elastic fibres are found, for the most part, within the opaque masses described. But, by the use of strong acetic acid, we often succeed in rendering these masses quite transparent, even though they be of considerable thickness ; when the enclosed elastic fibres, which resist, as we know, the action of this acid, become beautifully apparent. The tyro sometimes regards small fragments of cotton-fibre, etc., which likewise resist the action of acetic acid, as objects of suspicion. Needle-shaped crystals of fatty acids (sebacic, etc.) too, which often occur in abundance in the sputa of purulent bronchitis, pulmonary gangrene, and similar affections, may in their long, tortuous forms give rise to some un- certainty ; but, when slightly warmed, these melt, and are resolved into small fat-globules. One should make it a rule to diagnose elastic fibres as such, only when several of these lie together, and show plainly, by their character- istic, tortuous appearance, that they formed a portion of an alveolar wall. Larger pieces of the pulmonary paren- chyma, comprising several alveoli, are occasionally found. The larger and the more abundant these (microscopic) pulmonary shreds are, the more serious and rapid is the work of destruction which is going on.

It should moreover be observed, that sometimes in cases of pulmonary gangrene, elastic fibres can no longer be detected in the shreds of tissue which are coughed up, and recognized under the microscope as being of pulmon- ary origin ; inasmuch as they are gradually dissolved by putrid fluids. Traube, who first pointed out this fact, perceived in it a diagnostic criterion between gangrene and abscess of the lungs ; the elastic fibres being longer preserved in the contents of an abscess. In the ordinary caseation of lung tissue, also, the elastic fibres are lost ; though very gradually, it is true.

Fibrinous masses likewise occur in the sputa under certain conditions : coarse fibrinous casts of the larger bronchial tubes, in croupous bronchitis ; of the smaller

tubes, in acute pneumonia (Remak). Under the microscope, these dichotomously branched masses of fibrine,
which may be recognized with the naked eye, exhibit the
known construction,—a felty mass of fine fibres, which
swell up and disappear under the action of acetic acid,
and in whose interstices are stored round-cells in abundance, and masses of bacteria.

We have already briefly referred to *asthma crystals.*
These are very sharp-pointed octahedra, which are found
in large numbers, during the asthmatic seizure, in the
tough sputal masses previously described ; while, during
the intervals between the paroxysms, they are usually
wanting. But they have occasionally been met with in
the sputa of persons free from asthma, and, hence, are
not pathognomonic of this disease. They are quite
analogous to the crystals which have been found in the
spermatic fluid, medullary substance of bones, blood—
particularly in leucæmia,—besides various other localities,
and, according to Schreiner, they consist of the phosphate
of an organic base.

d. Bacteria, Bacilli of Tubercle, Micrococci of Pneumonia.

BACTERIA are present in large numbers in the sputum,
because of the admixture of the oral fluids ; but, besides
these, there occur forms which come from the respiratory
apparatus itself, as, for example, in putrid and diphtheritic

bronchitis, in purulent bronchial and tracheal catarrhs, etc. However, we are not yet able to see our way in the maze of differing forms. The reputed "discoveries" of the germs of whooping-cough and measles, of the micrococci of diphtheria, etc., in the sputa, have never possessed the slightest value in the eyes of those best fitted to judge; while the detection of the bacilli of tubercle in that fluid (Koch) is of the very highest importance. For, through Ehrlich's method of staining, it has become apparent *that the specific bacilli are almost never wanting in the expectoration of phthisical patients, and that, on the other hand, their presence affords an absolutely certain indication of the existence of tubercular phthisis.* The method of staining, which proves simultaneously the presence of the bacilli and their specific character, is the same as that already given for the demonstration of these organisms in sections of tubercular tissues. The sputum, spread in a very thin layer over the surface of a cover-glass, is dried, heated, and afterwards stained for twenty-four hours with a concentrated solution of gentian-violet, or fuchsine, in aniline water; that is, a saturated, filtered, aqueous solution of aniline. A short time suffices, in case the staining is performed at a raised temperature. Decolorization is effected by means of strong solutions of mineral acids; as twenty-per-cent. hydrochloric acid, or, what is best, a three-per-cent. alcoholic solution of nitric acid. All the other forms of bacteria (schizomycetes) present in the sputum, together with the mucus, etc., which were originally tinged as well as the tubercular bacilli, have now parted with their color, *and the latter alone appear as small, deeply-stained rods.* These specific bacilli are generally met with in the sputum in a state of lively sporulation, the spores appearing as bright spherules which occupy the whole width of the bacillus. When several spores are present in a bacillus, the latter may appear to be resolved into a row of granules.

In order to facilitate the focusing of the microscope, the remainder of the specimen may subsequently be stained by any process which may be preferred ; though, as a matter of course, the color selected should contrast as much as possible with that of the tubercular bacilli. In the majority of cases we find this double-staining un-necessary (see p. 102 *et seq.*).

Significance of the Tubercle-Bacilli.

Severe and Light Forms of Pulmonary Phthisis.

We desire to make some observations concerning the *diagnostic and prognostic significance attaching to the discovery of tubercle-bacilli in the sputum ;* and, in so doing, we cannot avoid a short digression into the field of the general pathology of phthisis.

Pulmonary phthisis is comprised in the domain of tuber-

culosis.   The separation of cheesy pneumonia and cheesy bronchitis from the strictly tubercular form of phthisis, as urged by Reinhardt and Virchow, is naturally to be supported on the ground of descriptive anatomy ; but the complete divorce of these processes must be abandoned, because of the histological conditions which obtain, but more especially from an etiological and practical standpoint.   A more accurate histological investigation disclosed the same elementary construction in cheesy pneumonia and bronchitis, as in the so-called genuine tubercles.   The author definitely proved this in the year 1873, in his report *"Ueber lokale Tuberculose"* (Volkmann's Samml. kl. Vorträge).   *In every case of phthisis, including the apparently non-tubercular forms, there are always present the characteristic, submiliary, non-vascular nodules, containing giant-cells, etc., which represent the chief type of tubercle, arrived at its complete histological development.*

These facts, thitherto unknown, must have greatly modified Virchow's theory of the duality of phthisis, which was already beset with objections.

The commonplace observation of the supervention of a tubercular pleuritis during the progress of a cheesy pneumonia, could not thereafter be regarded as the invasion of a new disease, as Virchow and his school taught ; but rather as the simple extension to the pleura of the same disease-process, which had before existed in the lung.

The simplicity and clearness of this view, when contrasted with the earlier theory, becomes immediately apparent, and yet a long time elapsed ere it obtained recognition.

Several years later, Charcot and his school defended the "*unité de la phthisie*" in a succession of works, founded upon the same observations ; while Rindfleisch entertained similar views. The subsequent experimental investigations of Tappeiner, Cohnheim, Salomonsen, and others, based on Villemin's discovery of the inoculability of tuberculosis, and, particularly, the epoch-making discoveries of Koch, proved with the greatest certainty, that one and the same etiological entity lies at the root of phthisis and tuberculosis ; though, naturally, the part played by other factors, under some circumstances, is not thereby ignored.

Pulmonary consumption, therefore, is now to be indeed regarded as a tuberculosis of the lungs, which, in the majority of cases, is localized.

If, now, we discover in the sputum that parasite, which we know to produce tuberculosis, we are forced to conclude that a tubercular process is going on somewhere in the respiratory apparatus, including the mucous membrane of the mouth and pharynx.

Shall we further conclude from this that the individual in question is a victim of general tuberculosis, and, hence, doomed ? No ; that would be a gross error.

The bacillus of tuberculosis is present not only in those cases in which the disease is progressing more or less rapidly, and eventually becomes generally disseminated in the organism through the medium of the blood- and lymph-vessels ; but, also, where for years and decades it remains localized, and, it may be, finally heals completely—*i. e.* in cases of "local tuberculosis." The human organism deports itself very differently, in this respect, from that of Guinea-pigs and rabbits, which animals are principally employed in these experiments. If a tubercular affection occurs anywhere in these animals,—in the eyeball, for example, after inoculation of the anterior chamber,—it extends, apparently without exception, to the various organs of the animal in the course of a few weeks or months, and generally causes speedy death. Dogs are not affected in a similar way ; though still fewer data, respecting the effects of such inoculation upon this animal, are available. *On the other hand, it is perfectly certain that in man the affection caused by the tubercle-bacilli remains in many cases localized and relatively benign for years, and may, eventually, heal more or less completely.* It is true, however, that so long as the tubercular process exists, there is always danger of its suddenly assuming greater intensity, and exhibiting a rapid local or general diffusion without any demonstrable cause. In the majority of cases, the human organism seems to afford the tubercular bacilli only a moderately favorable support ; so that they generally increase very sparingly. Should certain, hitherto unknown, conditions favor a *rapid* development of the parasites, the progress of the disease is correspondingly accelerated. Unfortunately, we have not mastered, as yet, these conditions, by which the increase of the parasite is, in many cases, retarded or prevented : should we accomplish this, we would immediately possess the key to the therapy of tuberculosis. We may indulge in the hope that even this high goal is not unattainable ; at all events, the approaches are smoothed.

If, therefore, we discover tubercle-bacilli in a specimen of sputum, we expose a tubercular process, which may *possibly* become eminently dangerous, by reason of extensive and rapid local ravages, and from transition to other organs. However, the possibility of a very chronic, bland course, and indeed of a cure, is likewise present.

The recognition of pulmonary tuberculosis will now be rendered feasible in many instances, where, formerly, it would have been nearly, or quite, impossible to make a positive diagnosis. The tubercle-bacilli are present in the greatest abundance on the surface of every phthisical cavity, even of the smallest size ; on ulcerative tubercular defects of the bronchi, etc. ; and, because of the sharp prominence afforded these by their specific reaction to certain staining procedures, they are found more easily and quickly in the sputum than the elastic fibres, which alone enabled us formerly to recognize the presence of a destructive process in the lungs. Accordingly, a careful examination of the sputa will enable us to recognize the numerous mild cases of phthisis, which run a favorable course, and which were generally regarded formerly as "suspicious" pulmonary catarrhs, bronchitis, etc., and even such cases as cause only the slightest subjective annoyance.

That these are of very common occurrence, one readily convinces himself at the post-mortem table. Carefully conducted necroscopies show that nearly fifty per cent. of all healthy, powerful individuals in adult life, who have perished from accident or acute disease, bear the traces and remains of destructive phthisical processes in the lungs,—either in the form of cheesy masses, often incrusted with calcareous material ; or, of cavities, bounded by hard cicatricial tissue. In many of these cases the disease had run a completely latent course, or nearly so ; at all events, in the majority no suspicion of a severe pulmonary disease had ever been aroused ; and yet, in every instance, the tubercular bacilli could at some time have been demonstrated in the sputa.

That a tubercular pleuritis, or even a fatal tubercular meningitis, etc., may suddenly be lighted up from phthisis,—it may be of quite limited extent,—which is to all appearances healing most favorably, or which has thus far pursued quite a latent course, is known by every practitioner, and is often enough confirmed by clinical and anatomical facts.

Accordingly, while the discovery of tubercle-bacilli in the sputa should always lead us to make a grave prognosis, it by no means justifies a positive prediction that the result will prove fatal. It is, indeed, known that even extensive phthisical ravages in the lungs may come to an end under favorable circumstances, and that every incipient phthisis need not lead to the destruction of that organ. But, at all events, the diagnosis of tubercular disease will have a determinative influence upon the patient's regimen, etc. It is altogether probable, therefore, that many lives may be preserved by thus recognizing the presence of tuberculosis in its early stages ; since the patients may be promptly subjected to such influences, climatic or otherwise, as experience has shown to operate favorably in incipient phthisis.

To what degree the extent of the phthisical process may be inferred from the number of tubercular bacilli found in the sputum, further investigations only will disclose.

On the other hand, the constant absence of tubercle-bacilli from the sputa, may be regarded as a certain sign that the destructive processes of tubercular phthisis are not then going on in the lungs. But, if elastic fibres are present in the sputa, while the specific bacilli are wanting, we must conclude that destructive changes of a different nature are occurring ; such as the formation of abscesses, the disintegration of tumors, etc.

It remains to be said that certain chronic ulcerative processes occur in the lungs, which are not of a tubercular nature, and in which, accordingly, tubercle-bacilli are not found ; but these are extremely rare.   Such were the observations of Riegel in *diabetic* lesions of the lungs.   In the majority of cases of diabetic phthisis, however, bacilli are present in large numbers.   The anatomical appearance of the lungs in this disease, where the bacilli are wanting, may very closely resemble ordinary tubercular phthisis.

One can readily perceive, that the examination of the sputum for the bacilli of tubercle should be undertaken only when one is armed with the best instruments.   It is true that the bacilli may be discerned even with the weaker objectives, in favorable preparations ; and, should they be very numerous, the naked eye may at times suffice for their recognition after they have been stained. Yet, it would be quite unsatisfactory to employ any save

the best immersion lenses in this work ; as it is quite
possible, according to the repeated experience of the
author, for the bacilli, which may be present in the
preparation, to escape notice and be overlooked with the
weaker systems—dry objectives, for instance ; while, with
a sufficiently high power—six hundred diameters at least,—
they become plainly apparent. At all events, it is impossible to say positively that tubercular bacilli are *not* present,
except one employs a strong immersion lens of excellent
quality, preferably homogeneous, and, as a matter of
course, Abbe's illuminating apparatus. He who fears
the cost of these expensive accessories, or shrinks from
the inconvenience attending their use, must not undertake examinations for the detection of bacteria (schizomycetes).

MICROCOCCI OF PNEUMONIA.

As we know, a peculiar structure, a capsular formation,
has been found investing the micrococci of acute croupous
pneumonia, which is plainly perceptible when a dry
preparation is stained with gentian-violet or fuchsine.
By this process the capsule takes a fainter stain than the
coccus proper, though its border is generally well defined.
Whether this structural peculiarity will aid us often in
forming a diagnosis, is still a matter of doubt ; though
several observers have testified with apparent positiveness
in the affirmative. However, capsulated micrococci

sometimes occur in the sputum even when pneumonia is not present, so that caution must be exercised. At all events, *colorless spaces* are often observed about micrococci, which signify very little ; while the micrococci of pneumonia possess capsules *which may be stained.*

3.—Pus.

Pus consists in general of a fluid, the liquor puris (Eiterserum), which holds in suspension small round cells, or pus-corpuscles ; and, in the majority of cases, micro-organisms also.

a. Pus-Corpuscles and Fat-Granule Cells.

The pus-corpuscles are very similar to, if not identical with, the white blood-globules, lymph-corpuscles, etc. When examined in a fresh state, that is in pus which has been secreted only a short time previously, they exhibit amœboid movements, and have then the characteristic, shining appearance of living protoplasm. In most cases, however, they have already perished, and their protoplasm is then coagulated in coarse granules. The nucleus, or nuclei, is usually hidden. One often sees very fine fat-granules scattered through the protoplasm, and these are abundant in pus-globules which have already been lifeless for a considerable time, particularly, therefore, in the pus-cells of so-called cold abscesses.

As a general thing, the pus-globules are nearly uniform

in size, about corresponding to the medium white blood-corpuscles. Frequently, however, there are mixed with these, larger cells, usually with vesicle-like nuclei. Should numerous granules of fat collect in these, the familiar *fat-granule cells* are formed, which, too, are sometimes still living, and exhibit amœboid movements. They appear, even under a low power, as dark little heaps in the midst of the ordinary pus-cells, exceeding the latter in size. Their dark color, when viewed through an objective of low power, not infrequently misleads the beginner to think of pigmentation ; though it is due solely to the large number of fat-granules, disposed one above the other. The light, which enters from below, passing so frequently from the fluid into the fat-granules, and from these again into the fluid, or the protoplasmic substance, is reflected as from spherical mirrors, and does not enter the eye of the observer ; hence the impression of darkness. On the other hand, these same fat-granule cells naturally appear of a bright white color, when we illuminate them from above, in consequence of the numerous reflecting surfaces. If, then, we cut off the light from the mirror, the fat-granule cells appear like brilliant white spherules in a dark field, provided that the light can reach the specimen from above. The high-power objectives, however, must usually be approached so closely as to shade the preparation completely.

However, real pigment-granules, the remains of blood-extravasations, are sometimes present in pus, being for the most part included in cells.   They are characterized by a brownish color.   Sometimes hæmatoidine crystals are also present.

b. **Other Ingredients.**

Pus is secreted either from a free surface, as a mucous or serous membrane, an ulcerous area, etc., or from within the tissues ; and, in both cases, it frequently contains substances derived from its place of origin.   Now, inasmuch as the source of the purulent discharge is, in many cases, unknown, though a point of great moment, it becomes clear, *that these adventitious ingredients of the pus assume a high diagnostic value.*

Thus, the important question whether an abscess has any connection with *bone*, is quite frequently determined by an accurate microscopical examination of the pus. The irregularly formed fragments of bone, which are found in the pus in these cases, often with sinuate resorption surfaces, the so-called Howship lacunæ, afford absolutely certain proof.   They are clearly distinguished by the marked brightness of their calcareous ground-substance, as well as by the characteristic star—or spider-shaped bone corpuscles, and may be found even with a lens of low power.   When necessary, the preparation may be cleared with a solution of caustic potash, this agent causing the pus-cells to disappear. In case the pus is not too thick, it may be allowed to stand until a sediment appears, when this is examined with the microscope.

Again, we find constituents of the food, from which we conclude that a communication with the alimentary tract exists. Or epithelium, elements of tumors, etc., appear, which often clear up the diagnosis in a surprising manner, and exercise a direct influence upon the treatment of the case. To cite an example: In the pus of a puerperal abscess, situated in the neighborhood of the thyroid gland, there appeared flat, cornified epidermic cells in large numbers, besides an abundance of cholesterine. A suppurating *branchial cyst* was immediately diagnosed, and the sac was extirpated. Echinococcal cysts, too, which have undergone suppuration, are often recognized as such only after a microscopical examination of the pus; in the course of which, either whole scolices, the characteristic hooklets, or the lamellated homogeneous membranes, have been found.

c. Schizomycetes and Actinomycetes.

In the pus of acute abscesses,[*] micrococci are present in great numbers, arranged in chain-like rows for the most part, and visible either in the fresh state, without

---

[*] Siehe Ogston, über Abscesse. *Arch. f. klin. Chir.*, Bd. 25.

further preparation, or after the familiar staining process has been employed.

In cases of chronic suppuration, the presence of micro-organisms in the pus is inconstant. Naturally, this is true only in those instances where the external air does not come in contact with the pus; otherwise, of course, saprophytic micro-organisms are always present.

In tubercular suppurations (peri-articular abscesses and suppuration of joint cavities in arthritis fungosa, scrofulous and carious abscesses, cheesy suppuration of lymphatic glands, etc.), tubercle-bacilli are naturally of frequent occurrence, and are always pathognomonic of this condition when found; but they are not present as regularly as in pulmonary phthisis. In many genuine tubercular abscesses, the bacilli cannot be demonstrated; indeed, according to Schlegtendal (*Fortschr. d. Med.,* 1883, S. 537), this is true of about half the cases.

Less frequently, other organisms are present in pus, apart from those which subsequently develop in a purely accidental manner; as, for example, those micrococci which cause the coloration of "blue pus."

The *actinomycetes*, which were first discovered in man by Langenbeck, later by J. Israël, Ponfick, and others; in beeves, by Bollinger,* may be recognized macroscopically in the pus, as jelly-like granules of the size of millet-

---

* Vgl. Ponfick, die Actinomycose, Berlin, 1882.

grains. When the granules are crushed, their peculiar structure appears ; staining is superfluous.

4.—URINE.

In the microscopical examination of the formed elements of urine, there first come into consideration :

a. Sediments and Crystals.

The *urates*, chiefly the urate of sodium, which are normally held in solution at the temperature of the body, are precipitated, when present in any excess, in the form of fine, and often somewhat irregularly formed, granules, when the urine cools. The author has quite frequently known these to be regarded by inexperienced observers as micrococci, and their trembling molecular motion as a vital phenomenon. A slight warming suffices to redissolve the urates, likewise the addition of acids ; when, in the latter instance, the uric acid separates in the form of characteristic crystals, prismatic for the most part, and frequently of a brownish color.

In feverish states of the system, gout, etc., the urate of soda is usually markedly increased ; and the separation of the uric acid often occurs spontaneously in the urine, some time after its evacuation, *i. e.* without the addition of acids. This was formerly regarded as "acid fermentation" of the urine, but wrongly so. The acid phosphate of soda is transformed into the basic form by the decomposition of the urate of soda, the uric acid being liberated. Acid fermentation proper of the urine, with an increasing acid reaction, occurs only in diabetes (Voit and Hofmann).

Simultaneously with the appearance of the uric acid crystals, a separation of the *oxalate of lime*, in the form of small, glistening octahedra, often occurs, which, when viewed in their shorter diameter, appear like square letter-envelopes (Briefcouverts). An abnormal abundance of oxalates in the urine, constitutes the condition known, as "oxaluria." The oxalate of lime, as is well known, forms the principal constituent of an important class of urinary calculi.

Later, the *alkaline or ammoniacal fermentation* regularly occurs in the urine. The urea is converted into the carbonate of ammonium, through the agency of an amorphous ferment, which was first isolated by Musculus. This ferment, however, is always produced by bacteria. Should these organisms, or their germs, find access to the urine in the bladder,—and this is mostly effected through the agency of the catheter and other urethral instruments, —the alkaline fermentation may be set up within the bladder itself, and particularly so when the urine stagnates in that viscus, as in vesical paralysis, etc.

During the alkaline fermentation the urine becomes markedly cloudy, this condition being caused not only by the presence of bacteria in swarms, but by the following substances :

1. Ammonio-magnesian phosphate, or triple phosphate, —a crystalline deposit, of which the typical form is a triangular prism with bevelled ends ("coffin-lid" crystals), very characteristic and easily recognized ; acids cause their immediate solution.

2. Urate of ammonium, in the form of dark spherules, the surface of which is either smooth or spiculate.

3. Phosphate of lime, which forms an amorphous sediment.

In pathological conditions we find, in addition to the substances already mentioned, which occur sometimes in enormous quantities, constituting the material known as "gravel" (Harngries), various other crystalline and granular deposits : as *cystine*, which is made up of regular six-sided tablets of different sizes (cystinuria) ; *xanthine* and allied bodies ; *tyrosine*, in the form of very fine needles, arranged in tufts, or "sheaf-like" collections, the latter often crossing each other and intersecting at their constricted central portions,—they are usually of a yellowish color, and occur principally in acute yellow atrophy of the liver ; phosphate and carbonate of lime, gypsum, etc. All these deposits are, as a general thing, easily detected by simple micro-chemical reactions, which are explained in the chemical manuals.

b. Urinary Casts.*

We have chiefly to distinguish between three kinds of casts : the *hyaline, waxy,* and *brown.* The hyaline casts consist of a perfectly transparent substance, very delicately contoured ; and hence they may easily be over-looked. Sometimes they are loaded with fat-granules, and are thus rendered more conspicuous. In breadth they often about correspond to the diameter of a red blood-corpuscle only, though they usually exceed this. We find them in the (albuminous) urine during the most varied pathological processes, as well as in cases in which neither inflammatory nor other like conditions prevail in the kidneys ; in many febrile states ; in icterus, etc. They are accordingly to be regarded as accompaniments of every, even the slightest, albuminuria.

On the other hand, the *waxy* casts, when found in con-siderable numbers, are of no inconsiderable diagnostic importance. They are always to be looked upon as sure signs of a kidney affection,—whether it be congestion, or a nephritis proper.

The waxy, or colloid, casts are composed of a sub-stance having more marked contours, and a somewhat glistening appearance, resembling that of molten wax ; or

---

* These were discovered in the urine by Henle, whose anatomical studies enabled him to recognize them immediately as casts of the renal tubules.—*Zeitschr. f. rat. Med.*, Bd. I.

they are slightly dulled by the presence of granular matter, usually the granular débris of a degenerated epithelial lining of a tubule, or of blood-corpuscles. In the
latter case they have even been distinguished as a particular variety, under the name of *granular* casts.

In diameter they vary greatly, reaching 0.05 mm. ($\frac{1}{500}$
of an inch) and over ; while in form they are usually
perfectly round, with a circular cross-section ; though
sometimes, and particularly in acute nephritis, possessing
irregular eroded borders. Very often these waxy casts
contain small round cells, or minute drops of fat ; while
occasionally they are covered with epithelial cells. Indeed,
there are casts consisting almost wholly of epithelial cells,
fused together more or less firmly. These are called
"epithelial" casts. The term "fibrine casts," or "cylin-
ders," which was formerly in vogue, has rightly been
quite abandoned. The substance composing the casts is
essentially different from fibrine, since acetic acid neither
dissolves it nor causes it to swell ; merely the glistening
appearance of the casts, their dark contour, and eventually their granular matter, are wont to disappear in this
medium, whereby they come to resemble the pale, hyaline
variety.

The casts are stained a light yellow by iodine ; indeed,
the waxy variety, in many cases, assumes a color ranging
from a dark yellow to a reddish brown.

We cannot, in the limited space of this work, enter upon the controversies concerning the origin of casts, simply mentioning the fact that the hyaline variety seems to be a direct product of exudation, while the waxy casts might be produced, in part at least, from lifeless epithelial cells.

Riedel * has discovered peculiar *slender brownish casts* in the urine of patients suffering from bony fractures, the casts being present for a few days after the reception of the injury. He asserts, and it is probably true, that these are products of the fibrine ferment, which has found entrance into the circulation. In these cases one frequently finds in addition, larger or smaller quantities of fat, which collects in the form of minute drops in the uppermost strata of the urine. This fat has entered the circulating blood at the point of injury, to be later arrested as emboli in the renal vessels, and thence gradually eliminated with the urine. In renal hemorrhages, also, and in hemorrhagic nephritis, casts often appear which are of a brown color, tinged by altered hematine ; though real blood-casts, too, are frequently observed.

c. Mucus- and Pus-Corpuscles. Epithelial Cells.

Lymphoid cells, mucus- and pus-corpuscles often occur in the urine. They may come from the kidneys, urinary

---

* Riedel : über das Verhalten des Urins bei Knochenbrüchen, *Dtsch. Zeitschr. f. Chir.*, Bd. 10.

passages, or from an abscess which has opened into the urinary tract. Epithelial cells, too, are often present, whose place of origin cannot always be determined with certainty. In catarrhal conditions of the urethra, or bladder, epithelial cells are sometimes found inclosing one or more lymphoid cells.

These latter were formerly considered to be of endogenous origin, but are now universally regarded as *invaginated* (Volkmann und Steudener)—that is, as having subsequently made their way into the interior of the epithelial cells.

Fat-granule cells occur only rarely in the urine ; Leyden found them in acute nephritis.

d. Tumor Components.

These are not difficult to diagnosticate in the urine, when some care is exercised ; though, of course, one must first become familiar with the various epithelial forms which are present in the kidneys and urinary passages. The multiform vesical epithelial cells in particular, which are frequently of considerable size and provided with several nuclei, have often been regarded as cancer-cells.

Diphtheritic, and cheesy tubercular, masses may also occur in the urine. They generally come from the bladder.

e. Entozoa.

These are very seldom present in the urine : echino-cocci, filariæ—the latter having as yet been found only in the tropics, in cases of chyluria,—and the eggs of *Distoma hæmatobium.* Many errors of observation have been recorded in this connection. One author, for example, described the ova of *Stongylus gigas,* which he had observed in a specimen of urine. A closer investigation disclosed the fact, that he had been deceived by granules of lycopodium, which formed an accidental impurity in the preparations.

f. Vegetable Parasites.

The *Sarcina urinæ* has been found in the urine, though very seldom.

As for *bacteria* and *micrococci,* their occurrence in the urine should naturally be determined by an examination, while the fluid is in an absolutely fresh state. *Normal urine is always free from such organisms ;* though, in a few hours after its evacuation, it may contain them in vast numbers. In fresh urine, as was first shown by Traube, these micro-organisms occur in the greatest numbers, when, after catheterization, ammoniacal fermentation of this fluid takes place in the bladder, lighting up a cystitis. In these cases we find, besides an abundance of triple phosphate, lymphoid cells, etc.,—vast swarms of bacilli and micrococci, which are deeply stained by the basic aniline dyes, and are often massed together in large heaps,—the *zoöglœa form*. The bacteria, whose germs have been introduced into the bladder by the catheter, frequently find their way through the ureters into the pelves of the kidneys and the renal parenchyma proper, and are found in the contents of pyelonephritic abscesses (Klebs).

In various infectious diseases, and particularly in metastatic suppurations, the organisms pass from the blood into the urine, as may assuredly be proved ; though, as yet, few reliable clinical studies on this subject have become known.

Those forms of bacteria which supply the ferment causing decomposition of the urea, never, so far as is known, pass from the blood into the renal secretion (probably because they are not present in it), but are always introduced from without. On the other hand, they must be present in the alimentary canal ; for the urea which enters the intestines (in cases of uræmia) is very speedily transformed into the carbonate of ammonium.

The specific micrococci which are present in the urine during gonorrhœal cystitis, do not decompose the urea,

In tuberculosis of the kidneys and urinary passages, I have in many instances demonstrated the tubercle-bacilli : at first on the cadaver only ; but later I have found them repeatedly in the living subject, and have utilized the fact in forming a diagnosis. The prognosis, in most cases of tubercular disease of the urinary apparatus, is very unfavorable.

5.—Secretions of the Genital Apparatus.

a. Vaginal Secretion.

The secretion of the vagina is a fluid containing more or less numerous large, imperfectly cornified, flat epithelial cells, besides round cells. The latter vary in size from that of the white blood-corpuscles to forms four and five times as large. Most of the larger round cells contain fat-granules in abundance.

Micro-organisms occur in large numbers in the secretion of this organ ; since all the conditions, requisite for an abundant development of injurious and harmless parasites, are afforded here, as well as in the oral cavity.

Among the innocuous parasites may certainly be reckoned the *Trichomonas vaginalis,* an infusorium provided with flagella and cilia, and capable of rapid motion, which was discovered in the vaginal mucus by Donné.

Mould-fungi (Schimmelpilze) also appear on the vaginal mucous membrane, particularly during pregnancy ; and, if developed in abundance, they form whitish patches, and occasion a slight catarrhal inflammation. According to Haussmann, this is the *Oidium albicans,* or thrush-fungus ; and, by infection with it, the thrush of new-born children may be produced. The numerous forms of cleft-fungi (schizomycetes) occurring in the vaginal secretion, cannot, as yet, be identified with certainty.

Concerning gonorrhœal micrococci, see sect. *c.* Of these microbes, also, too little is known to warrant us in founding a diagnosis upon their alleged appearance, in doubtful cases.

b. Uterine Fluids.

Dysmenorrhœal Membranes ; Decidual Shreds ; Diagnosis of Uterine Carcinoma.

Besides the normal plug of mucus in the cervical canal, which contains only a few lymphoid cells, we find in inflammatory conditions a fluid, and often pus-like, secretion of the uterine mucous membrane, in which numerous columnar cells, for the most part without ciliated borders, occur, in addition to lymphoid cells.

The *menstrual fluid* consists chiefly of blood ; while the *lochial secretion* contains in addition numerous elements from the remains of the oval membranes, and particularly from the deepest decidual layer, which *post partum* remains behind in the uterus. Especially characteristic are the epithelial cells lining the deepest portions of the uterine follicles, with their bright, almost vacuole-like, nucleus.

If portions of the placenta have been retained, shreds of various sizes are often found in the lochia ; and so characteristic is the structure and form of the arborescent *placental* (chorionic) *tufts*, that they are always immediately recognized as such. Calcareous incrustation of such placental remains often ensues ; and these firm concrements are at times expelled from the uterine cavity. A histological examination of these at once discloses their origin.

In certain forms of dysmenorrhœa, as is well known, membranes are expelled, the process being often accompanied with labor-like pains. An examination of these dysmenorrhœal membranes regularly discloses the fact, that they represent portions of the uterine mucous membrane itself. The tubular glands, and their orifices on the inner surface of the exfoliated tissue, may also be made out. Formerly, such membranes were generally denominated " *deciduæ menstruales* "; and the question, whether such cases did not really indicate an abortion during the early weeks of pregnancy, was discussed. Whether this be the case or not, a microscopical examination decides ; for the histological structure of the decidua of pregnancy—*decidua graviditatis*,—that is of the uterine mucous membrane as altered by conception, is very evidently different, and quite characteristic of gestation. According to our present knowledge, indeed, the interglandular tissue of the uterine mucous membrane consists, under all circumstances, save pregnancy,—whether it be during the menstrual tumefaction, in the dysmenorrhœal membranes attending the various forms of endometritis, when swollen in consequence of uterine myomata, etc.,—of *small* round-cells, similar to lymphoid elements in size, and possessing very little protoplasm.

The gravid state alone, effects a characteristic alteration in the cells. In the very first stages of pregnancy, we find in the swollen uterine mucous membrane the familiar large *"decidual cells," from five to ten times larger than lymphoid cells, and provided with abundant protoplasm.* These are round or polygonal in shape, and are also provided with prolongations. The cells of the decidua then maintain this size and form till the end of pregnancy. Except for the glands, the tissue of the decidua resembles, in some measure, many forms of large-celled sarcomata.

In extra-uterine pregnancy, also, a tumefaction of the endometrium regularly occurs, as we know ; and shreds of this membrane are frequently expelled. Here, too, the characteristic structure of the decidua, the large cells, may always be recognized (Wyder).*

An examination of membranes extruded from the uterus, therefore, enables us to diagnose, or to exclude,

---

* Wyder: *Arch. f. Gynäkologie*, Bd. 13.

pregnancy with certainty.   Moericke has shown that the above also applies to portions of mucous membrane removed from the uterine cavity with the sharp spoon.*

CARCINOMA, ADENOMA OR EROSION.

In cases of *uterine carcinoma*, cellular elements, or even larger fragments and shreds, are often held suspended in the fluid which is secreted from the ulcerating surface, and their microscopical structure tends to confirm the diagnosis.   In doubtful cases, however, where the question whether it be carcinoma or benign erosion is concerned, the examination of the secretion alone will never suffice ; in such instances, small pieces of tissue are often excised, upon the histological examination of which the diagnosis is to be founded.   Because of the practical importance of this question, we shall here insert a few observations concerning it.

The base of an *erosion* is made up of granulation-tissue, which is usually invested with several laminar coatings of epithelial cells.   From this epithelial invest-ment, hollow, gland-like prolongations, often, too, *solid columns of epithelial cells*, penetrate the granulation-tissue, where they ramify, the branches uniting to form an irregular network.

As one instantly perceives, this is a structure exactly

---

* Moericke : *Zeitschr. f. Gynäk.*, Bd. 7.

resembling that of carcinoma, and yet it occurs in quite simple, benign erosions. An incautious observer, who, from such an appearance of things, declares the case to be one of cancer without further ado, may easily cause great distress ; he will perform mutilating and dangerous operations, in cases where a radical extirpation is not indicated. Before making the serious diagnosis of cancer, therefore, still another point must be considered. The *exuberant epithelial growth*,* which penetrates secondarily the granulation-tissue, and which, frequently enough, becomes completely *atypical*, is not limited to uterine erosions, but appears frequently and in the most various localities, as the skin, liver, lungs, etc. It may occur in any place where granulation-tissue comes into direct connection with epithelial surfaces. The atypical growth of epithelium is not in any way altered by the character of the original affection, which led to the formation of the granulation-tissue. It is a perfectly benign, harmless process, and, from a practical standpoint, would interest us but slightly, were it not for the fact that the structure of cancers, in their early stages especially, bears the completest resemblance to these benign, atypical epithelial proliferations.

Indeed, from a histological point of view, cancer has

---

* Vgl. C. Friedländer, über Epithelwucherung und Krebs, Strassburg, 1877.

been defined as "atypical epithelial proliferation" (Waldeyer). However, this definition does not suffice; we must add, as the necessary and most important attribute of cancer : "*of a malignant character.*" In this we abandon the field of pure histology; for the "malignant character" can be directly recognized, neither from the cell, nor from the tissue. Nevertheless, a microscopical examination once more serves our ends; *for the malignancy of the process is certain, if it be found to pursue its course, untrammelled and destructive, through different tissues*, since a benign neoplasm remains limited to the histological formation in which it originated, leaving the neighboring structures quite untouched, or simply displaced.

If in the uterus, for instance, we find that the disease is not confined to the mucous membrane, but that it has invaded the muscular tissue as well; if we find the muscular structures partially replaced by granulation-tissue, through which atypical bands of epithelium ramify,* the malignancy of the growth becomes evident, and we are justified in pronouncing it to be a case of cancer.

The pieces removed for histological examination must, therefore, include a portion, at least, of the muscular structure; and, except it be established that the disease is

---

* The stroma of young, cancerous growths is usually granulation-tissue.

present in this tissue, we are not safe, in the majority of cases, in pronouncing it carcinoma.

In this matter I stand opposed to C. Ruge,* who proposes a very simple mode of distinguishing between cancerous ulceration and benign erosion.

Ruge asserts that the epithelial offshoots are solid in carcinoma ; hollow, or possessed of lumina, in simple erosions ; and, in many instances, this is certainly true. If, however, he means that this distinction obtains *without exception*, and may therefore be employed as a diagnostic, he is completely in error ; for, on the one hand, in many cases of genuine malignant carcinoma, the cellular rootlets possess the most beautiful, regular lumina ; while, on the other, solid epithelial offshoots are very frequently observed in benign erosions. That this is true, one may easily convince himself, particularly by observations on the cadaver. Ruge's criterion for the diagnosis of cancer, must accordingly be rejected, as completely untrustworthy.

The same principles apply to all other organs, as well as to the uterus, when this question is to be decided. Whenever we discover atypical epithelial neoplasms in localities where epithelium previously existed, as in the cutis, mucous membranes, all true glands, etc., we shall always be furthermore obliged to adduce the special

---

* C. Ruge : *Berl. klin. Wochenschr.*, 1878.

proof of malignancy, before making a diagnosis of cancer. If, however, we find the atypical epithelial growths
elsewhere,—in the muscular tissue,* bone, or lymphatic
glands, for instance,—the phagedenic, or metastatic,
character of the affection is thereby disclosed, and we
necessarily regard all such as cases of carcinoma.

Many authors use the term *epithelioma* synonymously
with *cancer;* but, in my opinion, this is not to be recommended. We would retain the word "epithelioma" as
a general name for *epithelial tumors of all kinds*, the expression "cancer" being applied to such of these as are
of a malignant nature. The term "adenoma" is best
employed for benign growths solely.

As yet, we know nothing at all of the reason why
cancers are malignant ; and still, on the ground of
a thousand-fold experience, we are enabled to make a
lethal prognosis with great positiveness in every case
where, in accordance with the principles just enumerated,
"cancer" must be diagnosed. If total extirpation of the
growth is not effected, the patient perishes in a very short
time,—within a few years. Only very seldom are there
exceptions to this empirical rule ; as, for example, cer-

---

* In teratoid neoplasms, epithelial structures may be found in the
midst of other tissues, though this fact does not bestow upon them
a malignant character. In uterine myomata, cysts have in rare
instances been found, which were lined with ciliated epithelium
(Babesin, Diesterweg).

tain flat forms of carcinoma cutis (ulcus rodens) occur, which may run a very protracted course.

On the other hand, one is enabled to exclude the presence of *immediate danger*, when the criteria of malignancy, as explained above, are wanting. That a cancer may at times develop from what was originally a benign growth, is naturally not impossible ; though positive danger ensues only when the malignant nature of the affection is shown.

Under some circumstances the diagnosis may become difficult, as when an 'atypical epithelial growth becomes associated, secondarily, with a process which is destructive in itself ; for example, with a syphilitic or tubercular ulceration. In such cases special caution is enjoined, and it is best to first await the effect of an energetic local treatment.

c. Secretion of Gonorrhœa.

Neisser has found a specific micrococcus in gonorrhœal pus,* which is characterized by its tendency to unite in small clusters, the individual granules being relatively at a considerable distance from each other They very frequently lie on the surface, and in the protoplasm, of the pus-corpuscles. The same micrococci are present in the secretions of gonorrhœal con-

---

* Neisser : *Med. Centralbl.*, 1879.

junctivitis. They are easily stained with basic aniline dyes in the usual manner.

As yet, however, the characteristics of the micrococci of gonorrhœa are not known with sufficient nicety. We cannot, with certainty, distinguish these microbes, when they occur mingled with other micrococci, as in the vaginal secretion, for example.

d. Seminal Fluid and Prostatic Secretion.

In the seminal fluid are found the characteristic spermatozoa, which are in lively vibratile motion, when normal semen is in a fresh state. When the secretion is in a dried condition, as in the seminal stains on linen, etc., the spermatozoa may still be demonstrated, as a usual thing, by macerating the stain in the chloride-of-sodium solution. In doubtful cases, if one should discover small shining bodies resembling the heads of the spermatozoa, and fine filaments similar to their tail-like appendages, but no complete organisms, he should refrain from giving a positive opinion ; inasmuch as small oval and filamentous bodies, of a most diverse origin, might easily appear in dried stains of any kind. Complete spermatosomata alone, possessing head and tail *in continuo*, should suffice for the positive diagnosis of a seminal stain.

If the spermatozoa are wanting in the seminal fluid, one should distinguish between a temporary and a permanent azoöspermia (*sterilitas masculina*). Should several ejaculations follow each other in rapid succession, the fluid, according to several observers, becomes quite free from spermatozoa, the secretion of the testicles being exhausted for the time being. The ejected matter is then contributed only by the seminal vesicles, prostate, etc.

The *prostatic fluid* is generally mixed with the sperma ; though, at times, it is separately evacuated during the act of defecation, etc., by the pressure exerted upon the prostate gland. In addition to the laminated amyloid bodies, this fluid often contains a large number of octahedral or lanciform crystals, which, like the asthma crystals in the bronchial secretion, represent the phosphate of an organic base—the so-called "Schreiner base." For a long time these have been known as spermatic crystals ; and Fürbringer * has shown that they are contributed by the spermatic fluid, and occasion the characteristic odor of the semen. They may often be demonstrated even in the freshly evacuated semen, or prostatic fluid, in other cases only after this has stood for some time ; though, when ammonium phosphate is added to the secretion, the crystals appear in large numbers.

6.—CONTENTS OF STOMACH AND INTESTINE.

For a long time, vomited matter and the alvine dejec-

---

* Fürbringer : *Zeitschr. f. klin. Med.*, Bd. 3.

tions have been submitted to microscopical examination and, in many instances, such investigations have yielded results of great importance, from a diagnostic and therapeutic point of view.

a. Food Remnants.

These are naturally present, in considerable abundance, always.   Striped muscular fibres are regularly found in the fæcal evacuations when a meat diet is enjoyed (Frerichs).   If vegetable food be used, the fæces contain cellulose ; though, when digestion is normal, starch is either entirely absent, or present only in quite minute amounts.   A large portion of the starch grains is taken up even in the stomach, and they then no longer strike a blue color with iodine, but rather a light-yellow.

The presence of starch, to any considerable extent, in the fæces, indicates a pathological condition, which is generally accompanied with diarrhœa.

Among the animal tissues which resist the digestive processes, are elastic fibres or bits of elastic tissue, which nearly always pass through the intestines unchanged ; shreds of tendon or fascia, not sufficiently softened by cooking ; horny epidermic tissue ; fragments of large arteries, etc.   And it is not only those who are mentally diseased, or markedly gluttonous individuals, in whose evacuations such masses are found, often in large quantities, but frequently quite rational beings as well.   These so-called "intestinal infarctions" at times cause both patient and physician considerable anxiety, though an examination with the microscope immediately dispels this feeling.

Undigested substances of a vegetable nature, also, are often passed ; and Virchow has called attention to the frequent occurrence of the cellular carpellary membranes of the orange (Apfelsinenschläuche) in the excrement.

b. Epithelial Cells, Mucus, Etc.

Epithelial cells from the mucous surfaces and glands of the alimentary canal are very frequently found in the contents of the stomach and intestine, usually, however, in a greatly altered state. No special significance attaches itself to such a discovery

An increased amount of mucus, which usually contains lymphoid round-cells and mucus-corpuscles, attests the presence of a gastric or intestinal catarrh. At times the evacuations contain, in addition to mucus, fibrinous masses, either in the form of membranes, *or of strands, whose arborescent ramifications have united to form reticular structures.* These are to be regarded as the products of a pseudo-membranous inflammation of the colon, which, as we know, sometimes pursues a retiform course. The evacuation of these masses, which may assume quite formidable dimensions, is at times accompanied with severe, labor-like pains. They usually contain mucous, in addition to fibrine, and, hence, are but partially dissolved in acetic acid. Besides these substances, only round cells, or their remains, are to be found in such fibrinous exudates.

c. Entozoa.

Several varieties of these occur in the intestinal canal, —many as unimportant parasites ; while others, again, are of very dangerous import. The parasites themselves, or portions of the same, are found in the alvine discharges ; but, when examining the dejections with the microscope, the *ova*, particularly, are not to be overlooked. Accurate descriptions and drawings of these are to be found in text-books on pathological anatomy, and in the familiar works of Leukart, Braune, and Davaine.

In the year 1880, Perroncito made the important discovery that the "tunnel disease," which decimated the St. Gothard laborers, is caused by the *Anchylomum duodenale* (or *Anguillula intestinalis* and *Pseudorhabditis stercoralis*). The peculiar ova of these parasites were found in the greatest abundance in the stools of the patients. In the so-called "miner's anæmia," also, which is prevalent in the mines of Hungary, he found the same parasites.

It is altogether likely that a careful examination of the

alvine discharges will serve to throw new light upon
other, hitherto mysterious, diseases also.

d. Vegetable Parasites.

Of these, one of the most interesting is the *Sarcina
ventriculi*, a member of the group of schizomycetes,
which consists of regular cubical aggregations of roundish cells, arranged in series of four, or its multiples. It
has no clinical importance. Micrococci, bacilli, etc.,
occur in the stomach, though usually to no great extent ;
while the yeast fungus, *Cryptococcus cerevisiæ*, is often
present in abundance.

In the intestinal contents and fæcal evacuations, the
yeast plant occurs also ; but particularly micrococci and
bacilli, large and small, may be found in vast numbers.
Of the bacilli, certain forms, as the *Bacillus subtilis*
(Ferd. Cohn), are distinguished ; though, in this maze
of micro-organisms, specific pathogenic forms cannot as
yet be recognized, with the exception of the tubercle-
bacilli, which respond to a particular staining process.

Several organic forms, which take a blue color when
acted upon by iodine, are met with in the intestines. Of
special interest is the *Clostridium butyricum*(Prazmowsky),*
which is found in abundance in the lower segment of the

---

* Prazmowsky: " Untersuchungen über Entwickelungsgeschichte
und Formenentwickelung einiger Bacterienarten." Leipzig, 1880.

ileum and in the colon,—not in the upper portions of the intestinal canal,—particularly in connection with a vegetable diet (Nothnagel).*   In size these vary, as does the yeast fungus ; and their shape also differs.   They are either rod-like organisms, pointed at one or both ends, or of an elliptical, or more or less fusiform, appearance ; and frequently they are grouped together in chains or heaps. By the action of iodine, these are stained a blue color, either wholly or in their central portion only ; while, in the latter case, the poles, or even the complete periphery, receive but a yellowish stain.   They are probably identical with the *Bacillus amylobacter* (Van Tieghem), and perhaps represent the ferment which causes the butyric-acid fermentation of the intestinal contents.

Nothnagel has demonstrated still smaller organic forms in the fæces, which take a blue color with iodine.

In cases of tubercular ulceration of the intestines, the tubercle-bacilli may very often be found in the contents of the gut, and, accordingly, in the fæces.   Swarms of these specific microbes are regularly present on the surface of the ulcers, to aid, later, in the formation of a diagnosis, when the excrement is examined (Lichtheim und Giacomi, *Fortschr. d. Med.*, 1883, S. 3 und 150).   It should be remembered, however, that sputa, swallowed by phthisical subjects, may likewise account for the

* Nothnagel : *Zeitschr. f. klin. Medicin.*, Bd. 3.

appearance of the bacilli in the fæces ; but, in either case, their discovery certainly proves that tubercular disease exists somewhere in the organism. The other bacilli occurring in the intestines are immediately de-colorized by the acid treatment ; but certain forms of large, round "micrococci" appear in the fæces, which, like the tubercle-bacilli, resist the acid treatment and retain their color for a long time. Koch regards these round bodies as spores, which form exceptions to the general rule ; for, as we know, most of the spores of bacilli with which we are acquainted cannot be stained. However, they can never be confounded with the bacilli of tubercle, and thus occasion diagnostic errors, because of their spherical shape.

7.—EXUDATIONS, CONTENTS OF CYSTS.

The microscopical examination of exudative fluids, the contents of cysts, and similar collections, for diagnostic purposes, is very frequently undertaken in these days of exploratory punctures and incisions.

Simple as the technology of such investigations is,—confined as they usually are to the direct observation of the sedimentary deposits by the methods already de-scribed, or eventually to the staining of the dry prepara-tions,—it is, in many cases, exceedingly difficult to estimate correctly the diagnostic value of what we find.

Fat, in considerable amount, is often present in exudates, producing in these an opalescent or milky appearance. At times the fat is suspended in the fluid, as in the chyle, in the form of exceedingly minute, irregularly shaped, and dully glistening molecules, each of which probably represents a particle of oil, or fat, coated over with albumen; and hence it is not immediately recognized. If acetic acid or an alkali be added, however, these albuminous envelopes are dissolved ; and the fatty particles run together, forming larger, brightly glistening droplets. This so-called *Hydrops chylosus* always seems to occur only when an effusion of chyle into the abdominal or thoracic cavity takes place, in consequence either of traumatism, or of the damming-up of the chyle in the mesenteric vessels or thoracic duct.* In other cases, the fat comes from disintegrated fat-granule cells, and is from the beginning distinguished as such, by reason of its small, shining particles (*Hydrops adiposus*). This condition is observed in chronic inflammation, and carcinomatous degeneration, of the serous membranes.

Besides these, serous effusions also occur, which exhibit an opalescent or milky opacity, due solely to the presence of albuminous granules.

Red blood-corpuscles may be found in variable quanti-

---

* Vergl. Quincke, über fetthaltige Transsudate, *Deutsch. Arch. f. klin. Med.*, Bd. xvi.

ties ; often, too, in an altered state,—decolorized, shrunken, etc.

Lymphoid cells are scarcely ever wanting ; only they are far less numerous in simple transudates, than in exudates of an inflammatory nature. They often exhibit lively amœboid movements ; though in many cases, doubtless, they are already dead. Frequently these elements contain numerous fat-granules.

Endothelial cells may be found in the effusions of serous cavities, either singly or united together in laminæ. They, too, often contain fat, are frequently transformed into spherical bodies, possess one or more nuclei, and sometimes, too, contain vacuoles.

Epithelial cells—columnar, pavement, or spheroidal— occur in the contents of cysts, and often serve as clews to the real origin of the latter. Ciliated epithelial cells, also, are sometimes found, particularly in unilocular ovarian cysts, etc.

Tumor elements, when mingled with exudates, usually sink rapidly to the bottom, and, as a matter of fact, this takes place within the body ; hence, if such cavities be punctured high up, these structures may be missed altogether, though present in abundance lower down.

The carcinomata are most often concerned, the cells of which are characterized by their very unequal, and oftentimes large, size, by the magnitude of their nuclei, and by the diversity of their forms. Vacuoles, also, are very frequently present in these elements. However, the discovery of single or detached cells does not usually suffice for the diagnosis of cancer; and the beginner particularly had better suspend judgment, until he finds the cells *massed together in heaps or balls.* Quincke asserts, that the *glycogen reaction* can probably be utilized in forming a diagnosis, as it often succeeds in the case of cancer cells, while endothelial cells usually seem to contain no glycogen.*

Whoever desires to make diagnoses of this kind, must, above all things, *operate upon the cadaver,* familiarizing himself with the formed elements which occur in the various kinds of serous effusions; otherwise, errors are unavoidable. The diversity of form, exhibited by the endothelial cells and their offspring in simple subacute or chronic inflammations of the serous membranes, is often very surprising to the non-connoisseur, and may easily lead to a false diagnosis of cancer.

The *bacteria*, which are found in effusions, have as yet been studied but little; and it is possible, that a closer investigation of these organisms will yield results of diagnostic importance.

During the pleural and pericardial inflammations ac-

---

* Quincke : Ueber Ascites. *Deutsches Archiv für klinische Medicin.*, Bd. 30.

companying acute pneumonitis, the capsulated micrococci
of pneumonia are very often developed in great numbers, and their discovery may be of real service in clearing up perplexing cases. By puncturing the lung
substance during an attack of acute pneumonia, Günther
and Leyden have, in two instances, reported the discovery of micrococci in the alveolar exudates. Günther,
in his case, first observed the occurrence of "colorless
envelopes" about these specific micrococci.

Of animal parasites, the *echinococci* chiefly come into
consideration here ; and the chitine hooklets, and laminated membrane, safford certain microscopical proof of
their presence. A narrow inspection of the sediment,
even with the naked eye, often discloses, as small whitish
points, the scolices, either singly or in groups.

VII.—EXAMINATION OF THE SOLID STRUC-
TURES OF THE CADAVER, EXTIRPATED
TUMORS, ETC.

THE microscopical examination of cadaveric sol-
ids, of extirpated tumors, etc., is made in accordance
with the methods already described. The elements are
obtained in an isolated state in the juices, by scraping a
fresh surface ; or by teasing, if need be after previous
maceration. *For this purpose, a freshly-cut surface is*

always prepared with a perfectly clean knife ; otherwise one runs the risk of being continually disturbed by the presence of accidental impurities.

Sections of the fresh specimen, also, are made with the curved scissors, razor, double-knife, or freezing microtome, and are first examined in a weak chloride-of-sodium solution. Bismarck-brown and the iodine solution, principally, are used to stain these fresh sections.

Our objects should always be first examined in a fresh state, since this offers very many advantages. The substances exhibit their natural transparency ; and we are thus best enabled to trace the relationship existing between the histological structures which we find, and those differences which are apparent to the naked eye. Any *fatty degeneration*, which may be present, can be thoroughly observed in the fresh specimen only.

When demonstrative, elegant preparations, suitable for staining, are to be made, the *freezing microtome* should always be employed. In this way, and in a very short time, we are usually able to obtain completely satisfactory and faultless sections, which may be examined directly in a weak solution of sodium chloride. Subsequently, the nuclei may be beautifully stained with Bismarck-brown ; and, after a short bath in alcohol, the section is finally transferred to glycerine, or to oil of cloves and Canada balsam. Thus, in urgent cases, sections affording a complete view of the structures may be prepared from the fresh organ of a cadaver, or from parts removed from the living subject, in the course of a few minutes ; indeed, in the latter case, during the operation itself. Under some circumstances, this is very advantageous.

In many instances, however, the examination of the fresh preparation does not suffice. Large, and very fine, sections of fresh organs are wont to curl upon themselves, and are so very soft and fragile, that sometimes, in spite of the greatest care and the expenditure of considerable time, one does not entirely succeed in placing them, intact and well extended, upon the slide. When this is the case, the tissues must first be hardened. *Alcohol is the agent which is almost invariably employed for this purpose*, its effects being simple and easily controlled. Only in the case of the nervous system, and occasionally of the eyeball, is it necessary to abandon this hardening medium in favor of chromic acid and its salts (Müller's fluid).

It is best to place small pieces of the specimen in a large amount of absolute alcohol, when the concentrated fluid rapidly permeates the tissues, and, by its immediate coagulation of the albuminous substances, effects a prompt fixation of the structural elements (Page 34, *et seq.*).

When hardened, the preparations are always cut with the microtome (see p. 19, *et seq.*); since a large number of sections, succeeding each other in regular order, may thus be prepared in a short time, to be subsequently manipulated as one chooses. These are first examined in simple water and glycerine, after which the various reagents and staining fluids may be employed.

Oftentimes a single short examination is all that is necessary, it being our task merely to rubricate, or name, an existing process. Here it suffices to scrutinize a preparation, both in a natural state, and after it has been tinged with one of the nuclear stains. At other times, however, we meet with unsuspected, and often surprisingly new, structures and combinations ; when it is advisable to make a large number of sections, and preserve them in alcohol. New views and new questions frequently arise, it may be after the lapse of considerable time ; and these are verified, or disproved, by subjecting our preserved sections to some special treatment, which before had not been thought of.

As to the proper mode of procedure in individual cases, and the interpretation of the results for diagnostic purposes, and in the interests of pathological science, a special consideration of these points would cause us to exceed the bounds originally prescribed for this manual. For such work, one should possess both clinical knowledge, and a careful training in pathological anatomy or histology, together with great discretion ; and these can be obtained only by thorough service in this field. One should make it a cardinal rule, *never to regard a "find" as pathological until a direct comparison has been made with the normal organ, the treatment being the same in each case.* The non-observance of this principle, reasonable as it sounds, has led science into many great errors, and, in practical cases, is often accountable for the grossest blunders.

Oftentimes, too, new facts in normal histology have been discovered in the course of pathological investigations. The new finds were at first regarded as having a pathological significance ; and the attempt was made to construe them as the causes or products of definite diseases, until they were finally recognized as perfectly normal structures which had previously been overlooked. This was the case with the "fat-granule cells," for instance, which are found in the central nervous system of more advanced fetuses and new-born children. These were regarded as signs of an encephalitis, until it was afterwards shown by Jastrowitz and Flechsig that they pertain to the normal development of the white substance, representing intermediate links in that process.

Very frequently, in pathological discoveries, we have to do with *quantitative deviations from the normal ;* as in the proliferation of the nuclei, clouding of the protoplasm (which, even in the normal state, exhibits a certain grade of opacity), atrophy and diminution in the size of the cells. It is quite evident, that, in order to recognize gradual changes of this sort, a direct comparison with the corresponding normal structures is absolutely necessary, the treatment of the specimens being precisely identical.

Furthermore, we very often discover things which are not really normal, and yet possess almost no pathological value—subnormal finds. Chief among these are the many senile degenerative changes, which, as long as they do not occur in excess, are well-nigh indifferent. One should by all means notice these, without ascribing to them too much importance.

The pathological rank and clinical importance of a disease often depend largely upon its extent. The beginner is often disposed to attach too great a significance to his discoveries. For example, he regards as very extensive, an alteration which occupies the entire field of view in the microscope. Repeated observations alone can guard the tyro from faulty conclusions in this respect ; since it is only by examining different parts of the organ, and particularly, too, by the systematic use of low powers, that one gradually learns to judge correctly of the quantitative spread of the process throughout the organ. If,

for example, one finds several contracted glomeruli in a
kidney, he does not at once conclude that all are in that
condition ; but he first ascertains the relative percentage
of altered and normal Malpighian tufts ; whether the affection is a diffused one, *i. e.* extended throughout the
entire organ, or limited to certain points ; whether, in
the latter case, the foci are more or less numerous ; and,
finally, whether the interfocal structures have remained
in a normal condition, or are seemingly somewhat altered.
If a small portion only of the organ is involved, the affection may possess very little clinical import, though of
great intensity in that particular locality. It should be
remembered, that even the sudden removal of a whole
kidney, from an otherwise healthy organism, is borne
without any considerable disturbance.

On the other hand, an alteration, far less striking, but
prevalent throughout the entire organ, may be attended
with consequences very injurious to its function, and,
consequently, to the whole organism. This is the case
in glomerulo-nephritis, for example, an affection characterized by nuclear proliferation in the loops of the
Malpighian tufts, whereby these are rendered im-
permeable to blood. Thus the renal circulation is greatly
impeded, or even reduced to a minimum. The untrained
observer may very easily overlook this extremely important condition completely ; while the attention of the

connoisseur is immediately directed to it by the contrast afforded by the bloodless, but at the same time large, glomeruli, and the capillaries of the cortex and medulla, which are filled with blood.

One should always bear in mind the difficulties and complications which beset inquiries into the *origin of these abnormal processes.* We see only what *has* taken place, not the process itself ; and a direct conclusion concerning the latter, cannot be drawn from the former without further ado.   Since the motility of the cells, the migration of the white blood-globules, etc., have become known, the greatest possible caution has been exercised in this respect, particularly, as these questions are usually of no immediate practical interest.   Some twenty years ago, we supposed ourselves further advanced in this matter than we are at the present day.   In those days, even the beginner was always enjoined, in the case of tumors, for example, to first determine their "genesis." The purport of this request (taken in its true sense it was quite impossible) was, that the transitions of the diseased into the normal tissue were to be studied, and, even at the present time, investigations of this kind are called for in many cases.   However, one should not at once imagine, that because he has ascertained the transitional forms of the disease, he has also arrived at the history of its development ; for, in that case, as the repeated experience of the past teaches, the gravest errors would result.

The signs of malignancy in tumors, as well as the methods employed when searching for bacteria in sections, have already been considered.

www.ingramcontent.com/pod-product-compliance
Lightning Source LLC
Chambersburg PA
CBHW030641030726
47497CB00006B/1903